# BUILDING TRADES
# PRINTREADING - Part 1

## RESIDENTIAL CONSTRUCTION

### Second Edition

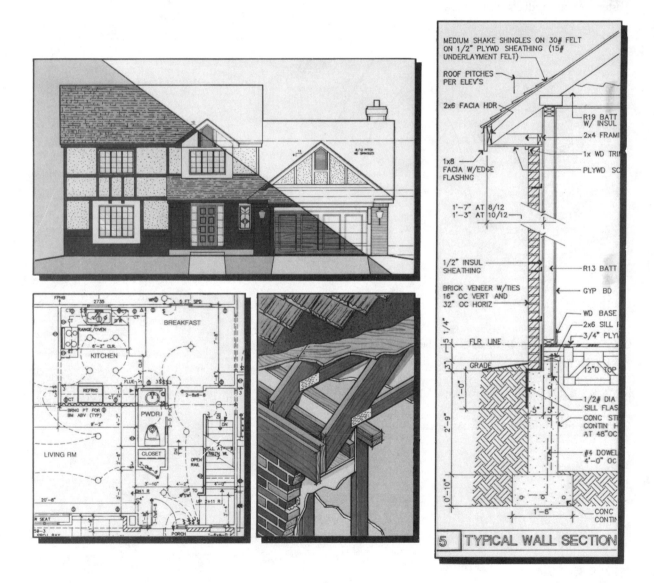

AMERICAN TECHNICAL PUBLISHERS, INC.
HOMEWOOD, ILLINOIS 60430

**Thomas E. Proctor**

## Acknowledgments

The author and publisher are grateful to the following companies and organizations for providing technical information and assistance.

The American Society of Mechanical Engineers
William Brazley & Associates
Rodger A. Brooks, Architect
Elkay Manufacturing Company
Hulen & Hulen Designs
Koh-I-Noor Rapidograph, Inc.
Western Wood Products Association

2 3 4 5 6 7 8 9 – 95 – 9 8 7 6 5 4 3 2 1

Printed in the United States of America

ISBN 0-8269-0443-2

# CONTENTS

**Working Drawings and Prints** — 1 — Working drawings, prints, and drafting methods are reviewed. — text... 1 — review... 13 — test... 17

**Working Drawing Concepts** — 2 — Pictorial and orthographic sketching and use of the architect's scale are presented. — text... 19 — sketching... 31 — review... 35 — test... 39

**Symbols and Abbreviations** — 3 — Symbols and abbreviations are shown and described. — text... 45 — sketching... 51 — review... 53 — test... 59

**Floor Plans** — 4 — Scale, orientation, and relationship of floor plans are studied. — text... 63 — sketching... 73 — review... 75 — test... 79

**Elevation Views** — 5 — Elevation and building design are studied. — text... 83 — sketching... 91 — review... 93 — test... 95

**Sectional Views** — 6 — Sectional views and residential construction are studied. — text... 99 — sketching...107 — review...109 — test...113

**Detail Views** — 7 — Detail views of structural and trim elements are studied. — text...115 — sketching...125 — review...127 — test...131

**Plot Plans** — 8 — Plot plans and survey plats are studied. — text...133 — sketching...139 — review...141 — test...143

**Trade Information** — 9 — Carpentry, masonry, electrical, plumbing, HVAC, and sheet metal work are related to printreading. — text...145 — review...167

**Printreading** — 10 — Final review based on text. Exams based on Stewart Residence Plans. — text...169 — final review...171 — final exam...177

Appendix...191

Glossary...207

Index...215

# INTRODUCTION

*Building Trades Printreading - Part 1,* 2nd Edition presents basic elements of printreading and provides printreading experience in residential construction. The text/workbook discusses conventional drafting, CAD, symbols and abbreviations, floor plans, elevation views, sectional views, detail views, and plot plans. Trade information, as related to printreading for residential construction, is also presented.

Two sets of plans are included in a separate folder as part of *Building Trades Printreading - Part 1,* 2nd Edition. Plans for the Wayne Residence are studied throughout the text/workbook. Plans for the Stewart Residence are used in Chapter 10 for the final exam. Additionally, one *trade plan* is included for study as assigned by your instructor. This plan may be used for additional printreading in your trade area.

Chapters 1 through 9 conclude with a variety of sketching exercises, review questions, and trade competency tests. Chapter 10 contains a final review exam and the final printreading exam. Specific instructions are given to complete each sketching exercise. Review questions and trade competency tests include True-False, Multiple Choice, Completion, Identification, Matching, and Printreading questions. Always record your answer in the space(s) provided. Answers for sketching exercises and all questions are in the Instructor's Guide for *Building Trades Printreading - Part 1,* 2nd Edition.

## True-False

*Circle T if the statement is true. Circle F is the statement is false.*

(T)   F      **18.** Dashed lines on floor plans show features above the cutting plane.

## Multiple Choice

*Select the response that correctly completes the statement. Write the appropriate letter in the space provided.*

___D___      **10.** Dimension lines may be terminated by _____.
                      A. arrowheads
                      B. slashes
                      C. dots
                      D. all of the above

## Completion

*Determine the response that correctly completes the statement. Write the appropriate response in the space provided.*

___R-19___      **18.** Batt insulation on the rear wall provides an insulation value of _____.

## Identification

*Select the response that correctly matches the given word(s). Write the appropriate letter in the space provided.*

___D___      **1.** Ceiling outlet

___B___      **2.** Bi-fold doors

___A___      **3.** Double-hung window

___C___      **4.** 240 V receptacle

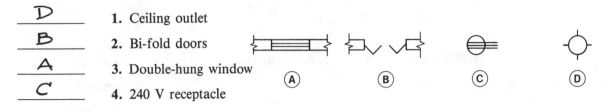

## Matching

*Select the response that correctly matches the given word(s). Write the appropriate letter in the space provided.*

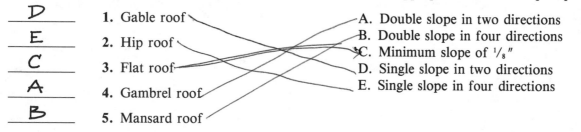

**D**    1. Gable roof      A. Double slope in two directions

**E**    2. Hip roof      B. Double slope in four directions

**C**    3. Flat roof      C. Minimum slope of $\frac{1}{8}''$

**A**    4. Gambrel roof      D. Single slope in two directions

**B**    5. Mansard roof      E. Single slope in four directions

## Printreading

*Study the referenced plan. Questions may be True-False, Multiple Choice, Completion, Matching, and so forth. Write the answer in the space(s) provided.*

**Refer to Torrance Residence on page 74.** — PAGE REFERENCE

**C**    1. A _____ house is shown.
- A. one-story
- B. one-story with basement
- C. one-and-one-half story
- D. two-story with basement

**11'-4"**
**23'-0"**    2. The living room measures _____ × _____.

**dashed**    3. The arch between the entry and the living room is shown with _____ lines.

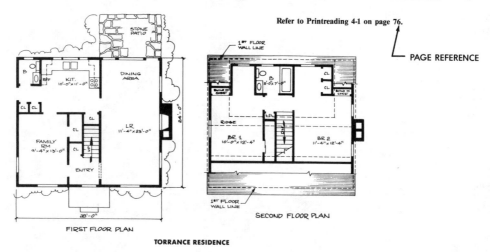

Refer to Printreading 4-1 on page 76.
— PAGE REFERENCE

FIRST FLOOR PLAN

SECOND FLOOR PLAN

**TORRANCE RESIDENCE**

## Plans

Two sets of plans are packaged separately from the book for convenience. Note that the size of each print has been modified and should not be scaled. Information along the border of each sheet identifies the print on each side of the sheet. For example,

- Wayne Residence—Sheet 1 (Sheet 2)
  - PRINT ON THIS SIDE
  - PRINT ON OPPOSITE SIDE
- Wayne Residence—Sheet 1 (Sheet 2)

An Index to Plans is located on the inside pocket of the folder containing the prints.

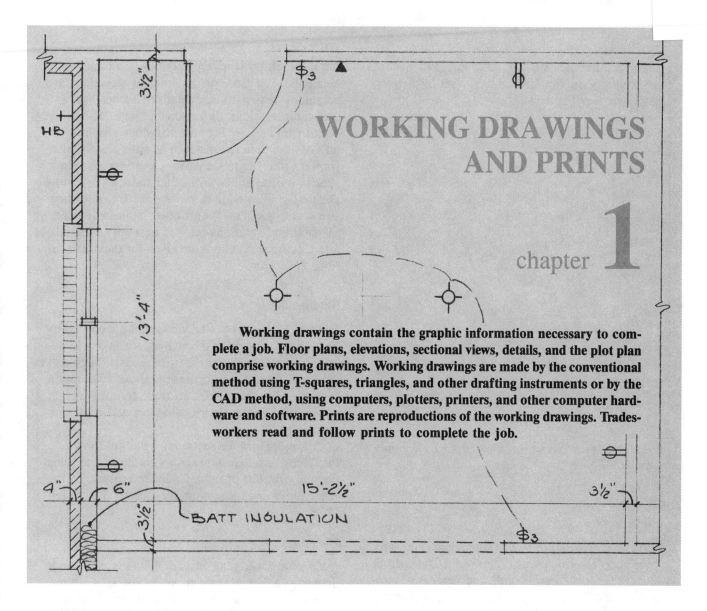

# WORKING DRAWINGS AND PRINTS

## chapter 1

**Working drawings contain the graphic information necessary to complete a job. Floor plans, elevations, sectional views, details, and the plot plan comprise working drawings. Working drawings are made by the conventional method using T-squares, triangles, and other drafting instruments or by the CAD method, using computers, plotters, printers, and other computer hardware and software. Prints are reproductions of the working drawings. Tradesworkers read and follow prints to complete the job.**

## WORKING DRAWINGS

*Working drawings* are sets of plans that contain the graphic information necessary to complete a job. *Specifications* supplement working drawings with written instructions giving additional building information. Working drawings consist of floor plans, elevations, sectional views, details, and the plot plan. A title block identifies each sheet.

Generally, the prospective owner (client) meets with the architect to discuss the planning of the house. The architect determines the client's lot size and shape. House styles are discussed, and the client's needs are determined, including desired number of rooms and their approximate sizes. Materials, equipment, and fixtures are discussed. The approximate price range of the house is determined, zoning requirements are discussed, and building codes are reviewed. With this information, the architect prepares a series of rough sketches for the client's review. The client makes the final decision, and the architect proceeds to complete the working drawings.

### Floor Plans

*Floor plans* are scaled views looking directly down on the dwelling. The cutting plane is taken 5'-0" above the finished floor to show the layout of rooms, information about windows, doors, cabinets, fixtures, and other features. The broad aspects of shape, size, and relationship of rooms are determined from the floor plan. Dimensions show sizes of rooms, hallways, wall thickness, and other measurements. Symbols and abbreviations give additional information about the rooms shown. A separate floor plan

1

is required for each story of the house. Floor plans are generally the first drawings of a set of plans to be drawn. See Figure 1-1.

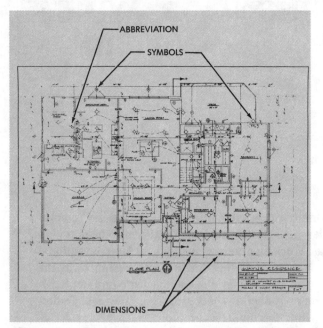

**Figure 1-1.** Floor plans show the size, shape, and relationship of rooms.

## Elevations

*Elevations* are scaled views looking directly at walls. They show the true shape of the walls. Dimensions include floor-to-floor heights and heights of windows above finished floors. The two types of elevations are exterior and interior elevations.

**Exterior Elevations.** *Exterior elevations* are scaled views that show the shape and size of the outside walls of the house and the roof. Four exterior elevations are generally required to show exterior walls. They are generally drawn to the same scale as the floor plan. Building materials such as brick, aluminum siding, or wood are specified for the exterior walls through the use of symbols and notations. Openings for doors and windows and their types and sizes are shown in their proper locations. The roof style, slope, and type of roof covering are shown.

**Interior Elevations.** *Interior elevations* are scaled views that show the shape and size of interior walls and partitions of the house. They may be drawn to

the same scale as the floor plan or to a larger scale when showing more detail. Interior elevations are commonly drawn to show details of kitchen base and wall cabinets, sinks, dishwashers, ranges, and other built-in appliances. They are also drawn to show details of bathroom fixtures such as water closets, tubs, showers, and vanities. Interior elevations also show special wall treatments in other rooms. For example, details of interior elevations are drawn to show fireplaces, bookcases, and other built-in features. Symbols and notations are used to specify materials, and specific instructions are given for their installation. See Figure 1-2.

## Sectional Views

*Sectional views* are scaled views that are created by passing a cutting plane through a portion of a building. Common sectional views are taken through outside walls to show information about foundation footings and walls, wall and floor framing, height of windows above floors, and eave and roof construction.

Cutting planes for sectional views are shown on the floor plans. Direction arrows on the cutting plane line show the line of sight. Coded references give the sheet number of the section drawing. See Figure 1-3.

## Details

*Details* are scaled plan, elevation, or sectional views drawn to a larger scale to show special features. Whenever a part of the building cannot be shown clearly at the small scale of the plan, elevation, or sectional views, it is redrawn at a much larger scale so that necessary information is shown properly. For example, a fireplace, framing for a stairwell, special interior or exterior trim, an entrance doorway, or a section of a foundation may be drawn as a detail. Sectional details show the cross-sectional shape of features such as foundation footings and windows. The scale for details is determined by the complexity of the detail shown. See Figure 1-4.

## Plot Plans

*Plot plans* are scaled views that show the shape and size of the building lot and the location, shape, and overall size of the house on the building lot. Plot plans are drawn to smaller scales than floor plans to show the full lot on a single sheet. Solid lines show

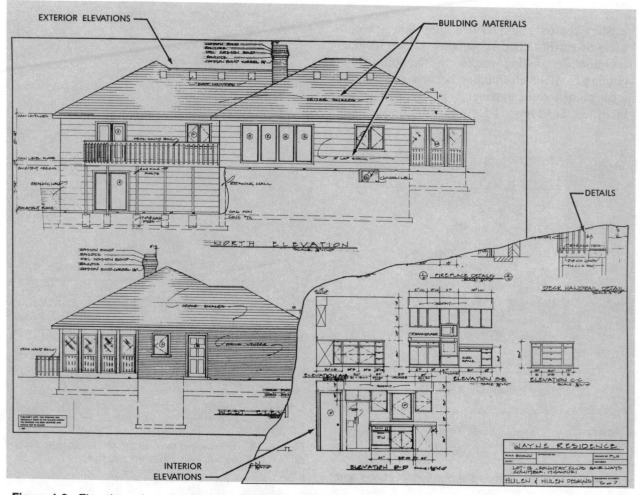

**Figure 1-2.** Elevations show the size and shape of exterior walls, interior walls, and partitions.

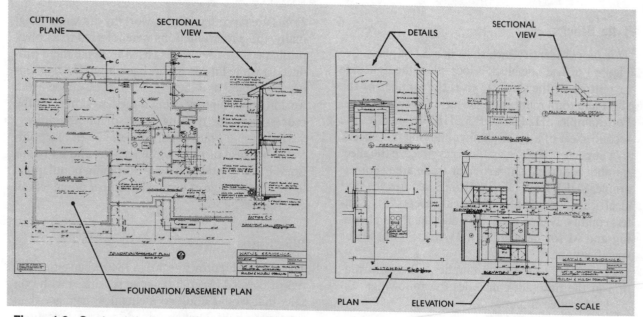

**Figure 1-3.** Sectional views show features revealed by a cutting plane.

**Figure 1-4.** Details show special features as elevation or sectional views.

existing contours, and dashed lines show new contours of earth. Other information shown includes a symbol designating true North, the point of beginning from which building corners and heights, location of streets, easements, and utilities are established. See Figure 1-5.

4. Date
5. Revisions
6. Drafter's initials
7. Checker's initials
8. Owner's name
9. Owner's address
10. Scale of drawings

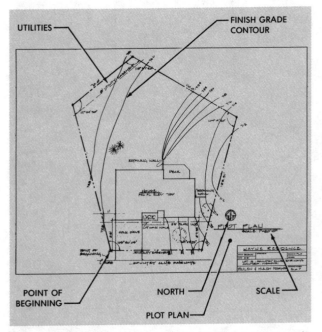

**Figure 1-5.** Plot plans show the building lot shape and size and location of the house.

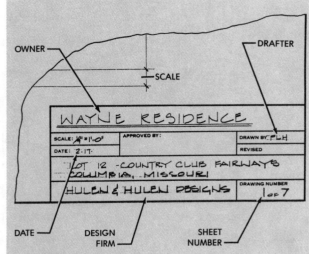

**Figure 1-6.** Title blocks include the sheet number, name of architect, owner, and other information.

## Title Blocks

*Title blocks* identify each sheet in a set of plans. See Figure 1-6. The title block is located on the right side or bottom of the sheet. It gives the number of the sheet and total number of sheets in the set of plans. For example, SHEET 2 OF 5 SHEETS denotes the second sheet of a set of plans containing five sheets. In a larger set of plans, initials may precede the sheet number to indicate a particular trade area. For example, SHEET E 1 OF 3 SHEETS denotes the first electrical sheet of three electrical sheets. Other letters commonly used to denote trades are P for plumbing and M for mechanical (HVAC). S is used for specifications.

In addition to the sheet number, other information in the title block includes the following:

1. Architect's name
2. Architect's seal
3. Project or building number

## PRINTS

*Prints* are reproductions of working drawings. Originally, these reproductions were referred to as blueprints because the process used to make them produced a white line on a blue background. Any number of copies or blueprints could be made from working drawings by using a process similar to the process used for making prints of photographs.

Diazo prints are generally preferred over blueprints today because of their white background and dark lines. Electrostatic prints are becoming popular due to their advantage of easy enlargement or reduction. See Figure 1-7.

## Blueprints

The use of blueprints began in 1840 when a method was discovered to produce paper sensitized with iron salts that would undergo a chemical change when exposed to light. Drawings made on translucent (allowing light to pass through) paper were placed over

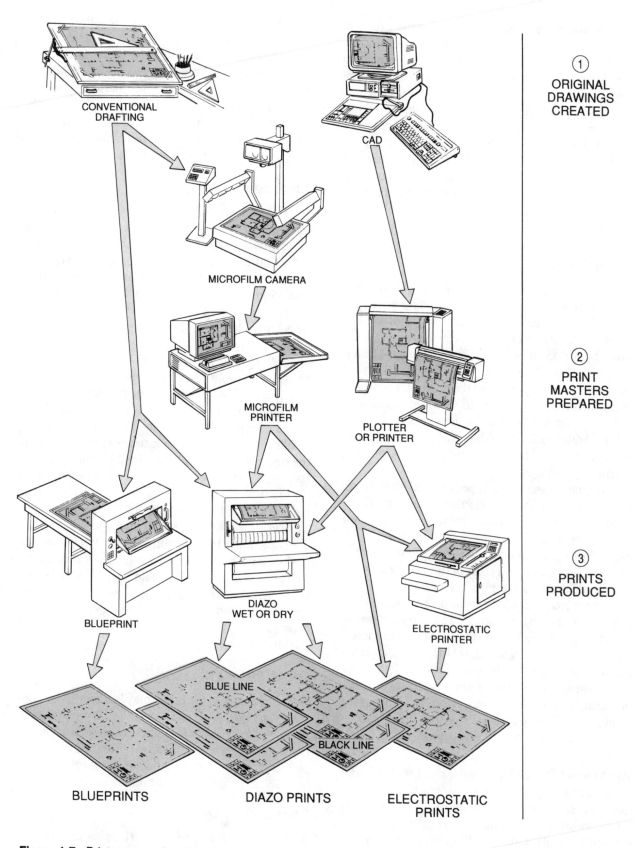

**Figure 1-7.** Prints are produced by the blueprint, diazo, or electrostatic process.

the sensitized paper in a glass frame used to hold the paper firmly. The frame was then exposed to sunlight. A chemical action occurred wherever the light was permitted to strike the paper. When the blueprint paper was washed in water, the part protected by the pencil or ink lines on the tracing would show as white lines on a blue background. A fixing bath of potassium dichromate, a second rinse with water, and print drying completed the process. Blueprints are not common today, although they are used in some industries such as oil production because they fade less rapidly in sunlight than prints produced by the diazo process.

## Diazo Prints

The majority of prints used today have blue or black lines on a white background. Blue line prints are generally preferred by engineers while black line prints are preferred by architects. These prints are made by the *diazo* process. This process has the advantage of providing excellent reproductions with very good accuracy because the paper has not been soaked with water and then dried. Diazo prints, with their white background, are easier to read than blueprints. Additionally, the white background provides a convenient area for writing field notes or making emergency changes.

Two types of sensitized paper are used in the diazo process, one for each development method. These papers are coated with a chemical that, when exposed to ultraviolet light, becomes a part of a dye complex. The original drawing or a copy, on translucent material, is placed over a sheet of the sensitized paper (yellow side) and is fed by a belt conveyor into the print machine. The two sheets revolve around a glass cylinder containing an ultraviolet lamp and are exposed to the light. The sensitized paper is exposed through the translucent original in the clear areas but not where lines or images block the light. The sheets are separated, the original is returned to the operator, and the sensitized paper is transported through the developing area. It is then developed by either a wet diazo or dry diazo process.

**Wet Method.** In the wet development method, the sensitized paper passes under a roller which moistens the exposed top surface completing the chemical reaction to bring out the image. Prints made by this method have black or blue lines on a white background.

**Dry Method.** In the dry development method, the sensitized paper is passed through a heated chamber in which its surface is exposed to ammonia vapor. The ammonia vapor precipitates the dye to bring out the image. Prints made by this method have black or blue lines on a white background. This method is most commonly used today since high-quality reproduction can be achieved on mylar (plastic film) or sepia (brown line) copies.

## Electrostatic Prints

Electrostatic prints are produced by the same process used by office copiers. Full-size working drawings are exposed to light and projected directly through a lens onto a negatively charged drum. They may also be photographed with a camera and reduced to a 35mm ($1\frac{3}{8}''$) frame of film. The film, after processing, is inserted into an aperture card which can be keypunched for computer sorting, filing, and retrieval. The microfilm in the aperature card is then exposed to light and projected through the lens onto the negatively charged drum.

The drum is discharged by the projected light from the nonimage areas but retains the negative charge in the unexposed areas. The drum then turns past a roller where black toner particles are attracted to the negatively charged image areas on the drum surface. As the drum continues to turn in synchronization with the positively charged copy paper, toner particles are attracted to the paper and fused to it by heat and pressure. See Figure 1-8.

Prints made by this method have black lines on a white background. The advantages of the electrostatic process include easy enlargement and reduction of drawings, small storage size, quick retrieval and duplication, and reduced shipping costs. The major disadvantage is the potential for distortion by projection through a lens.

## DRAFTING METHODS

Working drawings for prints may be made using *conventional drafting* practices or *computer-aided design*. A "language" of standard lines, symbols, and abbreviations is used in conjunction with drafting principles so that drawings are consistent and easy to read. See ANSI Y series for additional information.

## Conventional Drafting

Basic tools such as T-squares, triangles, drafting instruments, scales, and pencils are used to produce

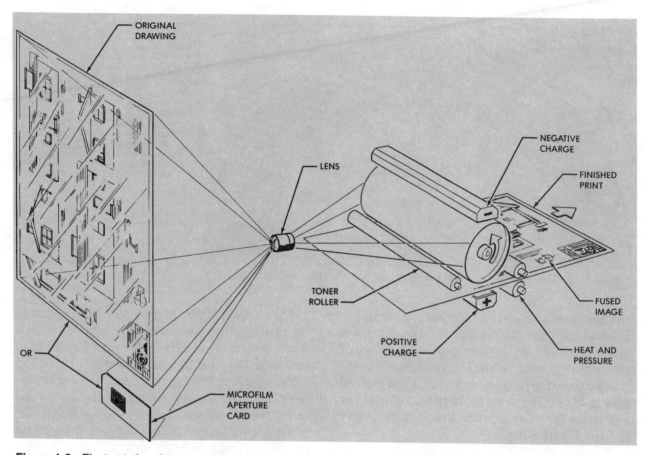

**Figure 1-8.** Electrostatic prints are produced as light is projected through a lens onto a negatively charged drum which offsets the image to positively charged paper.

working drawings by the conventional method. See Figure 1-9. Drafting machines (combination T-square, scale, and triangles) and parallel straightedges (combination drafting board and modified T-square) are commonly used in production situations. Drawings are begun after taping the drafting paper to the drafting board. All line work is done with construction lines that are darkened to produce the final drawing.

**T-squares.** The T-square is used to draw horizontal lines and as a reference base for positioning triangles. The head of the T-square is held firmly against one edge of the board to ensure accuracy. T-squares are made of wood, plastic, or aluminum and are available in various lengths. The most popular T-squares are 24″ to 36″ in length.

**Triangles.** Triangles are used to draw vertical and inclined lines. The base of the triangle is held firmly against the blade of the T-square to ensure accuracy. Two standard triangles, 30°-60° and 45°, are

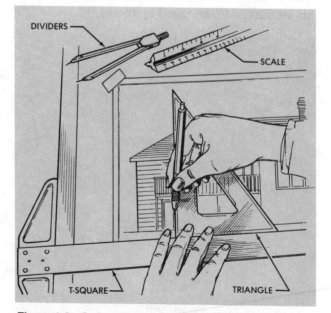

**Figure 1-9.** Conventional drafting tools include T-squares, triangles, instruments, scales, and pencils.

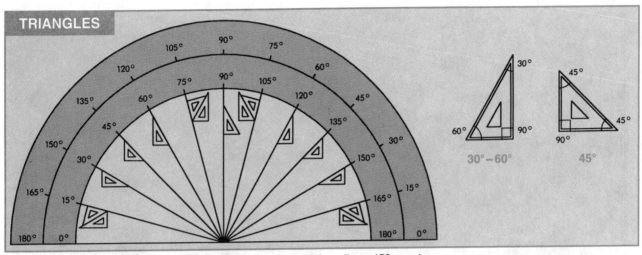

**Figure 1-10.** The 30°-60° and 45° triangles are used to draw lines 15° apart.

commercially available in a variety of sizes. Triangles are made of clear plastic. The 30°-60° triangle is used to produce vertical lines and inclined lines of 30° or 60° sloping to the left or right. The 45° triangle is used to produce vertical lines and inclined lines of 45° sloping to the left or right. The triangles may be used together to produce inclined lines every 15°. See Figure 1-10.

**Drafting Instruments.** Although a wide variety of precision drafting instruments is commercially available, the compass and dividers are the most commonly used. Each of these is commercially available in a variety of sizes. See Figure 1-11.

The compass is used to draw arcs and circles. One leg of the compass contains a needlepoint that is positioned on the centerpoint of the arc or circle to be drawn. The other leg contains the pencil lead used to draw the line. Two types of compasses are *center-wheel* and *friction*. The radius of the arc on the center-wheel compass is changed by adjusting the center wheel. Arcs of various radii are obtained on the friction compass by opening or closing the legs. Center-wheel compasses are the most popular and most accurate.

Dividers are used to transfer dimensions. Each leg contains a needlepoint to assure accuracy. Two types of dividers are center-wheel and friction. Friction dividers are the more useful for general work.

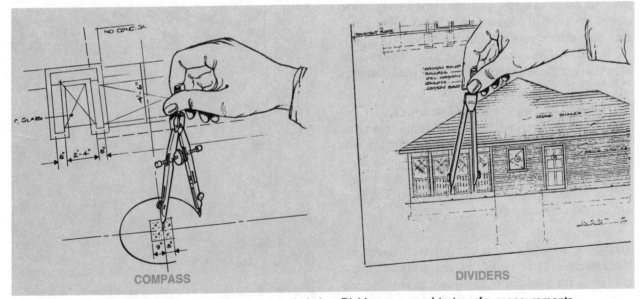

**Figure 1-11.** The compass is used to draw arcs and circles. Dividers are used to transfer measurements.

Other drafting instruments include irregular (french) curves and architectural templates. Irregular curves are used to draw curves that do not have consistent radii. Architectural templates are used to save time when drawing standard items such as doors, windows, and cabinets.

**Scales.** Scales are used to measure lines and reduce or enlarge them proportionally. The three types of scales are the *architect's scale, civil engineer's scale,* and *mechanical engineer's scale.* A variety of sizes is commercially available. See Figure 1-12.

An architect's scale is used when making drawings of buildings and other structural parts. A common

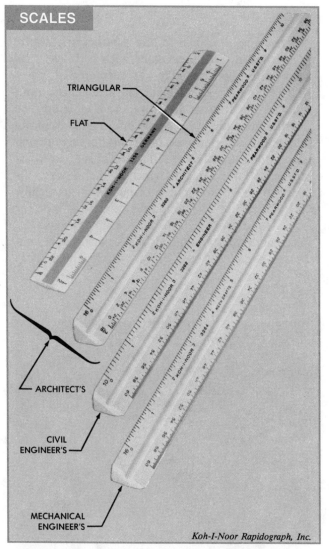

SCALES

TRIANGULAR

FLAT

ARCHITECT'S

CIVIL ENGINEER'S

MECHANICAL ENGINEER'S

*Koh-I-Noor Rapidograph, Inc.*

**Figure 1-12.** Three types of scales are used to produce scaled drawings. The architect's scale is used for building trades plans.

type is triangular in shape. One edge of the scale is a standard ruler divided into inches and sixteenths of an inch. The other edges contain 10 scales that are labeled 3, 1½, 1, ¾, ½, ⅜, ¼, 3/16, 3/32, and ⅛. The ¼ scale means that ¼″ = 1′-0″, and so forth. For larger scale drawings, the 1½″ = 1′-0″, or 3″ = 1′-0″ scales are used.

The civil engineer's scale is used when making maps and survey drawings. Plot plans also may be drawn using this scale. The civil engineer's scale is graduated in decimal units. One inch units on the scale are divided into 10, 20, 30, 40, 50, or 60 parts. These units are used to represent the desired measuring unit such as inches, feet, or miles. For example, a building lot line that is 100′-0″ long drawn with the 20 scale (1″ = 20′-0″) measures 5″ on the drawing.

The mechanical engineer's scale is used when drawing machines and machine parts. This scale is similar to the architect's scale except that the edges are limited to fractional scales of ⅛, ¼, ½, and 1 related to inches. Decimal scales are also available.

**Pencils.** Wooden or mechanical pencils are used to draw lines. Wooden pencils contain a stamp near one end indicating the hardness or softness of the lead. The lead of desired hardness is inserted into mechanical pencils.

Hard leads are used to draw fine, precise lines. Medium leads are used to draw object lines. Soft leads are used primarily for sketching. Grades of lead range from 6B (extremely soft) to 9H (exceptionally hard). The architect's range is HB, F, H, and 2H. F and H are most used for producing drawings. See Figure 1-13.

## Computer-Aided Design (CAD)

Computer-aided design (CAD) is also known as computer-aided drafting, or computer-aided drafting and design (CADD). This system is popular in architectural, and engineering offices. Advantages of CAD for architects and engineers include speed, accuracy, consistency, changeability, duplication, and storage. A major additional benefit is information management. As drawings are made or changed, information about the materials used in the drawings is stored and updated, allowing the operator to extract it for estimating or pricing purposes. Advantages of CAD for tradesworkers include legibility and

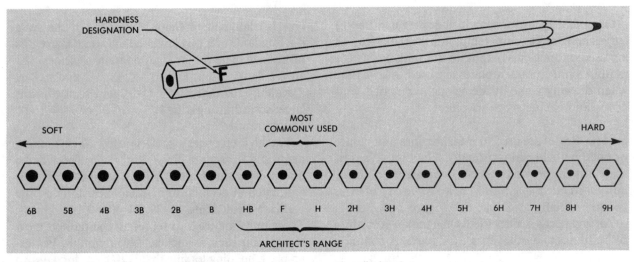

**Figure 1-13.** Pencil leads range from extremely soft to exceptionally hard.

consistency of line weights, notes, and symbols. Additionally, drawings may be printed in colors for greater clarity.

CAD systems use *hardware* (computing tools) and *software* (computing programs) to *input* (generate), *manage* (duplicate, file, and retrieve), and *output* (print or plot) drawings. A large variety of hardware components and software programs is commercially available for powerful mainframe systems and personal computer (PC) systems.

**Input Systems.** Input systems are composed of a software program such as AutoCAD® and the hardware such as the keyboard, central processing unit (CPU), and monitor. Other inputting hardware includes either a mouse, trackball, joystick, thumbwheel, tablet, or TouchPen™.

The keyboard is an input device similar to a typewriter keyboard with the addition of *function keys* (keys that perform special tasks) and a *numeric keypad* (numerical keys which may be locked ON for exclusive inputting of data). Drawings may be input with the keyboard; however this is not an efficient method.

The CPU receives information from the input device, processes it, and displays results on the monitor. A hard disk and large random access memory (RAM) are required for CAD programs.

The monitor is a video unit with high resolution allowing drawings, data, and text to be displayed. Monitors are either *monochrome* (1-color) or color. Color monitors are preferred because of their ability to distinguish layers. *Layering* is imposing one layer over another. This feature is particularly helpful when creating additional plans, such as the electrical plan, after the floor plan has been created.

The mouse and trackball are mechanical input drawing devices that are moved on a tabletop. The joystick and thumbwheel input information by movement of the stick or wheel in its own housing through electromechanical means. The tablet is a sensitized pad containing a menu area on which a stylus or a puck is moved to input information electronically. The TouchPen™ is pointed directly at the monitor screen to control the display. These input devices implement drawing commands that create, change, and move lines, arcs, circles, and solids. See Figure 1-14.

**Figure 1-14.** CAD input systems include the keyboard, CPU, monitor, and other input devices.

CAD drawings can easily be modified, rotated, revolved, or viewed from other angles. Additionally, they are easily copied and stored. Diskettes containing CAD drawings may be mailed, or files may be transmitted via telephone modem.

**Output Systems.** Output systems can display drawings through a monitor, print drawings on a printer, or plot drawings through an automated pen plotter. The method chosen depends on the size and complexity of drawing required.

Printers generate drawings by applying the image to the paper after receiving electronic information about the drawings from the computer. Print may be either on opaque paper, tracing vellum, or film and either black line or in several colors.

Dot matrix printers operate similar to a typewriter. A row of pins strikes a ribbon transferring carbon to the paper and creating the image. Laser printers function similar to an office copier. Information is electronically transmitted to the print paper which attracts toner particles forming the image of the drawing. The particles are then permanently fused to the paper by heat.

Plotters generate drawings with pens. The bed plotter retains the drawing paper in a stationary frame while a mechanism guides the pen across the surface in all directions, raising and lowering the pen to contact the surface and produce the drawing. The rotary-drum plotter moves the paper in one direction around the drum while the pen is raised and lowered and moved at a right angle across the drum. Lines at angles, curves, and circles are drawn by movement of both the paper and the pen. See Figure 1-15.

**CAD Plans.** Plans drawn by the CAD method are more consistent and standardized than plans drawn

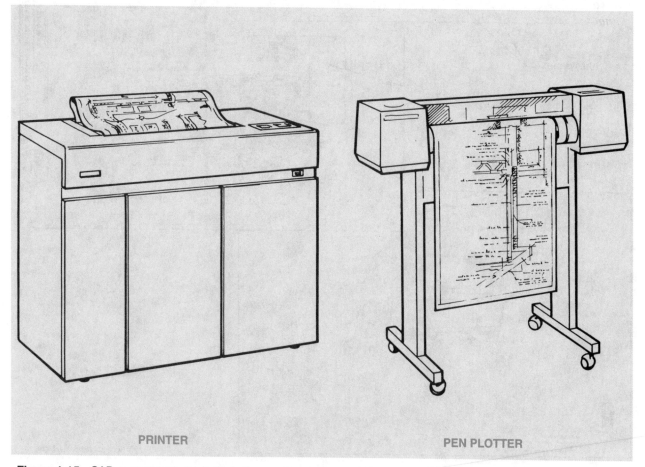

PRINTER  PEN PLOTTER

**Figure 1-15.** CAD output systems include the printer and pen plotter.

by the conventional method. Line weights, symbols, and lettering on CAD drawings are precise and easy to read. These same elements on plans drawn by the conventional method reflect the individual technique of the architect.

CAD plans are becoming more prevalent due to the lower prices of computer systems with sufficient capacity for CAD software programs. Architects are using CAD more because of the variety of software programs available and their ability to perform complicated tasks quickly. See Figure 1-16.

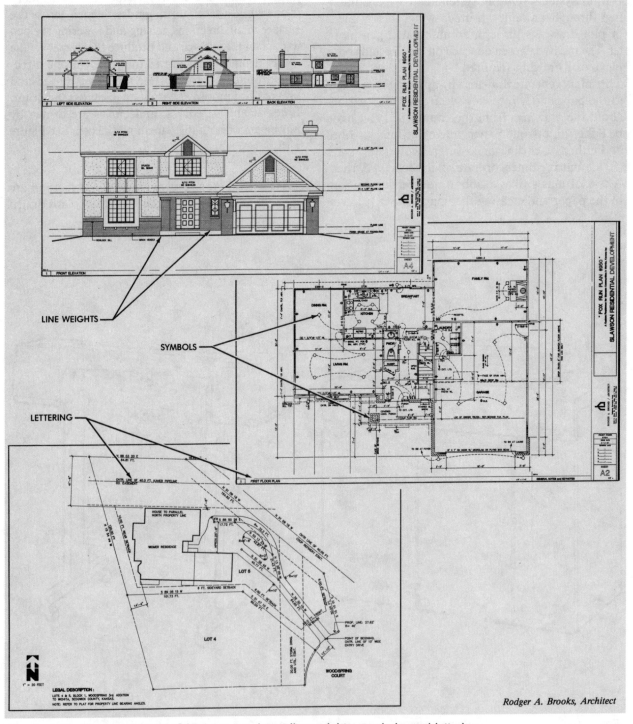

*Rodger A. Brooks, Architect*

**Figure 1-16.** Plans drawn with CAD have consistent line weights, symbols, and lettering.

# Review Questions

Name _____ Date _____

## Multiple Choice

_A_ 1. A true blueprint has _____ lines on a _____ background.
A. white, blue
B. white, black
C. blue, black
D. black, white

_D_ 2. The major disadvantage of electrostatic prints is their _____.
A. sensitivity to light
B. difficulty of duplication
C. large storage space requirements
D. potential for distortion during production

_B_ 3. Shape, size, and relationship of rooms is most clearly shown on _____.
A. plot plans
B. floor plans
C. elevations
D. details

_B_ 4. Translucent paper _____.
A. absorbs light
B. allows light to pass through
C. reflects light
D. prevents light from passing through

_D_ 5. T-squares are made of _____.
A. wood
B. aluminum
C. plastic
D. all of the above

_D_ 6. The most common diazo process utilizes _____ to produce prints.
A. silicates
B. peroxide
C. chlorine
D. ammonia

_C_ 7. The primary use of compasses on drawings is to draw _____.
A. horizontal and inclined lines
B. vertical and inclined lines
C. arcs and curves
D. none of the above

_____A_____    **8.** The _____ scale is preferred when drawing buildings and other structural parts.
     A. architect's
     B. civil engineer's
     C. mechanical engineer's
     D. drafter's

_____C_____    **9.** An F pencil lead is harder than a _____ pencil lead.
     A. 2H
     B. 4H
     C. 6B
     D. none of the above

_____D_____    **10.** Coded references on the cutting plane show the _____.
     A. scale of the details
     B. sheet number of the details
     C. both A and B
     D. neither A nor B

_____C_____    **11.** Two basic types of elevations are _____.
     A. rough and finished
     B. top and bottom
     C. interior and exterior
     D. scaled and freehand

_____C_____    **12.** Cutting planes for sectional views may be shown on the _____.
     A. plot plans and elevations
     B. plot plans and floor plans
     C. elevations and floor plans
     D. plot plans, elevations, and floor plans

_____B_____    **13.** The first drawing(s) of a set of plans to be drawn is generally the _____.
     A. elevations
     B. floor plans
     C. details
     D. plot plan

_____C_____    **14.** Cutting planes for floor plans are taken _____ above the finished floor.
     A. 3'-0"
     B. 4'-0"
     C. 5'-0"
     D. 6'-0"

_____A_____    **15.** The name and seal of the _____ are commonly found in the title block.
     A. architect
     B. drafter
     C. contractor
     D. owner

_____C_____    **16.** Regarding floor plans, _____.
     A. the layout of rooms is shown
     B. symbols and abbreviations give additional information
     C. both A and B
     D. neither A nor B

_B_  **17.** Regarding title blocks, _____.
  A. one title block is completed for each set of plans
  B. initials representing trade areas may precede sheet numbers
  C. both A and B
  D. neither A nor B

_D_  **18.** Regarding prints, _____.
  A. blue line prints may be made by the blueprint method
  B. blue line prints may be made by the electrostatic method
  C. both A and B
  D. neither A nor B

_B_  **19.** Regarding scales, _____.
  A. the architect's scale is graduated in decimal units
  B. the architect's scale contains a ruler and 10 scales
  C. the mechanical engineer's scale is graduated in decimal units
  D. none of the above

_C_  **20.** Regarding CAD, _____.
  A. the mouse and trackball are output systems
  B. the dot matrix printers are input systems
  C. plotters generate drawings with pens
  D. prints are initially produced on carbon paper

## True-False

T  (F)  **1.** Diazo prints fade less rapidly in sunlight than blueprints.

(T)  F  **2.** Electrostatic prints are produced by the same process used by office copiers.

T  (F)  **3.** Tradesworkers prepare working drawings of the final house design for approval by the owner.

(T)  F  **4.** Working drawings contain all graphic information necessary to complete a job.

(T)  F  **5.** Sectional views can show information about foundation footings and walls.

(T)  F  **6.** Elevations are generally drawn to the same scale as floor plans.

(T)  F  **7.** Each sheet of a set of working drawings contains a title block.

T  (F)  **8.** True South is designated on the plot plan to show house orientation on the lot.

(T)  F  **9.** Plot plans are drawn to a smaller scale than floor plans.

(T)  F  **10.** A separate floor plan is required for each story of a house.

(T)  F  **11.** Floor plans are generally the first drawings of a set of plans to be drawn.

(T)  F  **12.** The cutting plane for a sectional view is shown on the floor plan.

(T)  F  **13.** The scale of a print may be given in the title block.

(T)  (F)  **14.** A monochrome monitor can display up to 12 colors.

(T)  F  **15.** The most popular T-squares are 24″ to 36″ in length.

## Completion

*Prints*    1. _____ are reproductions of working drawings.

*machine*    2. The mechanical engineer's scale is used when drawing _____ parts.

*Plot*    3. The location of streets, easements, and utilities is shown on the _____ plan.

*head*    4. The _____ of the T-square should be held firmly against the edge of the drawing board when in use.

*Dividers*    5. _____ are used to transfer dimensions on drawings.

*Dimensions*    6. _____ on plans give the size of an object.

*Specifications*    7. _____ are written documents giving additional information about the plans.

*exterior*    8. The slope of the roof is shown on the _____ elevations.

*existing*    9. Solid lines on plot plans show _____ contours.

*draft*    10. A(n) _____ machine replaces the T-square and triangles in use.

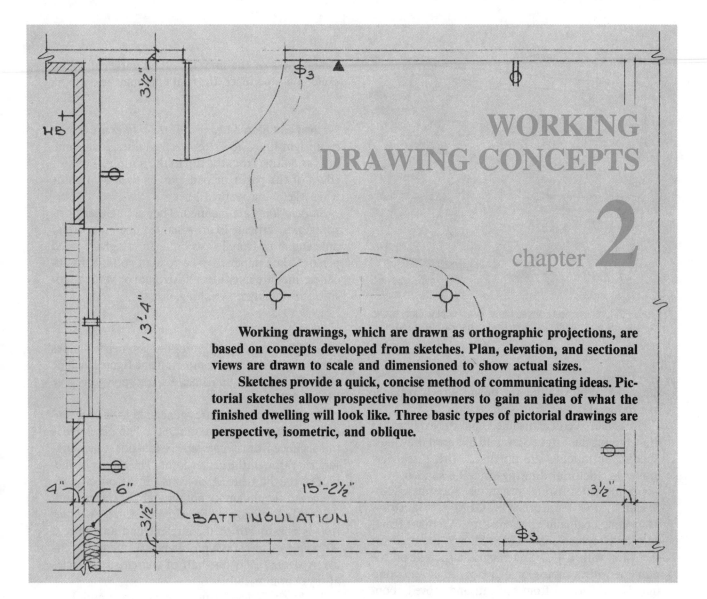

Working drawings, which are drawn as orthographic projections, are based on concepts developed from sketches. Plan, elevation, and sectional views are drawn to scale and dimensioned to show actual sizes.

Sketches provide a quick, concise method of communicating ideas. Pictorial sketches allow prospective homeowners to gain an idea of what the finished dwelling will look like. Three basic types of pictorial drawings are perspective, isometric, and oblique.

## SKETCHING

*Sketching* is drawing without instruments. Sketches are made by the freehand method. The only tools required are a pencil, paper, and eraser. See Figure 2-1.

Sketching pencils are either wooden or mechanical. Wooden pencils must be sharpened, and the lead must be pointed. One type of mechanical pencil contains a thick lead that is pointed with a file, sandpaper, or lead pointer. Another type contains a thin lead which does not require pointing. Softer leads, such as HB, F, and H, are commonly used for sketching.

Paper selected for sketching depends upon the end use of the sketch. Plain paper is commonly used. Tracing vellum is used if the sketch is to be duplicated on a diazo printer. Papers and tracing vellums are available in pads or sheets in standard sizes desig-

nated A, B, or C. Size A is 8½″ × 11″. Size B is 11″ × 17″. Size C is 17″ × 22″. The paper is either plain or preprinted with grids to facilitate sketching. Preprinted paper is available in a variety of grid sizes for orthographic and pictorial sketches. A grid size of ¼″ is common. Grids are commonly printed in light-blue, non-reproducing inks.

Erasers are designed for use with specific papers and leads. The eraser selected should be soft enough to remove pencil lines without smearing lines or damaging the paper. Pink pearl and white vinyl erasers are commonly used when sketching.

### Sketching Techniques

The pencil point should be pulled across the paper when sketching. Pushing the pencil point can tear

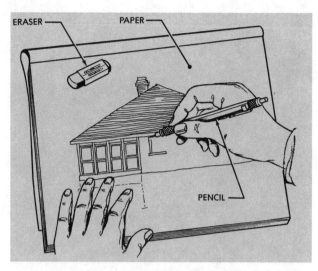

**Figure 2-1.** Sketches convey ideas graphically. Only basic tools such as pencils, paper, and erasers are needed.

the paper. While pulling the pencil, slowly rotate it to produce lines of consistent width.

Horizontal, vertical, inclined, and curved lines are drawn to produce three-view and pictorial drawings. Shading techniques are not used with three-view drawings. Pictorial drawings may be shaded.

Computer software programs such as Auto-Sketch™, PC Paintbrush™, GEM™, MacDraw, MacPaint, Logipaint™, Microsoft® Windows Paint, Mentor Graphics®, and Video Show™ are used for sketching with a PC. Sketching is displayed on the monitor with a mouse or digitizer. Lines and commands are selected from a menu and moved about the screen to create the sketch. Other tools used for PC sketching include a light pen, joystick, and keyboard. The sketches are printed on paper by any of the standard computer printing or plotting methods.

**Horizontal Lines.** *Horizontal lines* are level or parallel to the horizon. To sketch horizontal lines, locate the end points with dots to indicate the position and length of the line. For short lines, the end dots are connected with a smooth wrist movement from left to right (for a right-handed person). Long lines may require intermediate dots. If grid paper is used, intermediate dots are not required. For long lines, a full arm movement may be required to avoid making an arc.

The top or bottom edges of the paper or pad may be used as a guide when sketching horizontal lines. Light, trial lines are drawn first to establish the

straightness of the line. It is then darkened. With sketching experience, the trial lines may be omitted.

**Vertical and Slanted Lines.** *Vertical lines* are plumb or upright lines. To sketch vertical lines, locate end dots and draw from the top to the bottom. The side edges of the paper or pad may be used as a guide when sketching vertical lines.

*Slanted lines* are inclined. They are neither horizontal nor vertical. To draw slanted lines, locate end dots and draw from left to right (for a right-handed person). The paper may be rotated so that the inclined lines are in either a horizontal or vertical position to facilitate sketching.

**Plane Figures.** *Plane figures* are geometric shapes having a flat surface. Common plane figures include circles, triangles, quadrilaterals, and polygons. See Figure 2-2.

*Circles* are plane figures generated around a centerpoint. All circles contain 360°. The *diameter* is the distance from *circumference* (outside) to circumference through the centerpoint. The *radius* is one-half of the diameter. A *chord* is a line from circumference to circumference not through the centerpoint. An *arc* is a portion of the circumference. A *quadrant* is one-fourth of the circle. Quadrants have a 90° angle. A *sector* is a pie-shaped segment of a circle. A *semicircle* is one-half of a circle. Semicircles always contain 180°.

To sketch circles, locate the centerpoint and draw several intersecting diameter lines. Mark off the radius on these lines, and connect with a series of arcs. Darken the lines to produce a smooth circle.

*Triangles* are three-sided plane figures. All triangles contain 180°. *Right triangles* contain one 90° angle. *Obtuse triangles* include one angle greater than 90°. *Acute triangles* have no angles of 90° or more. Isosceles and equilateral triangles are acute triangles. *Isosceles triangles* contain two equal sides and two equal angles. *Equilateral triangles* contain three equal sides and three equal angles.

To sketch triangles, draw the base, determine the angle(s) of the sides, and draw straight lines to complete. Generally, one or more of the sides is dimensioned and the angle(s) noted.

*Quadrilaterals* are four-sided plane figures. All quadrilaterals contain 360°. *Squares* contain four equal sides and four 90° angles. *Rectangles* contain

## PLANE FIGURES

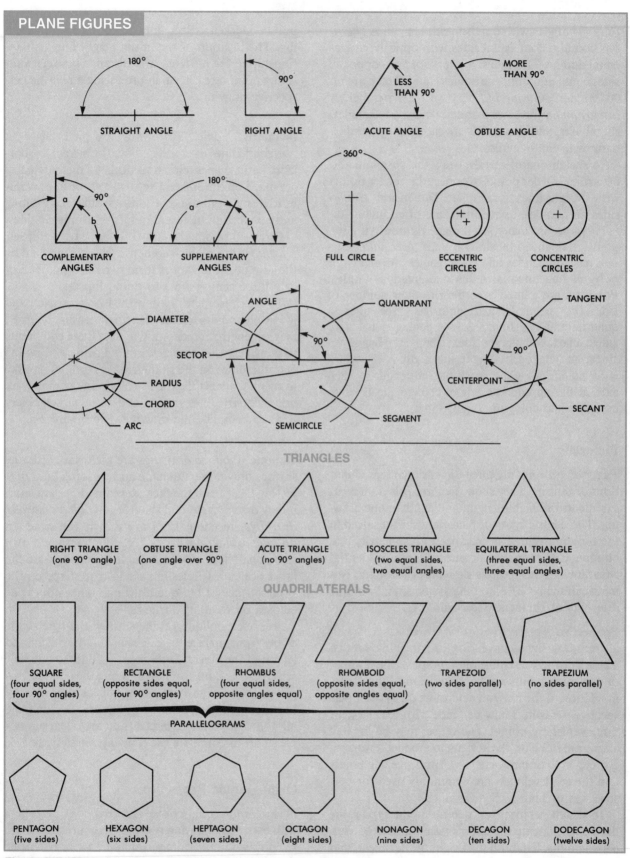

**Figure 2-2.** Plane figures are geometric shapes with flat surfaces.

four 90° angles with opposite sides equal. A *rhombus* contains four equal sides with opposite angles equal and no 90° angles. A *rhomboid* has opposite sides equal and opposite angles equal. There are no 90° angles. A *trapezoid* has two sides parallel. A *trapezium* has no sides parallel. Squares, rectangles, rhombuses, and rhomboids are classified as parallelograms in mathematics.

To sketch quadrilaterals, draw the base line and determine corner points. Connect the corner points with straight lines to complete. Dimensions of two sides and angles, as required, are often included.

*Polygons* are many-sided plane figures. All polygons are bounded by straight lines. A *regular polygon* has equal sides and equal angles. An *irregular polygon* has unequal sides and unequal angles. Polygons are named according to their number of sides. For example, a triangle has three sides; a quadrilateral has four sides; a pentagon has five sides; a hexagon has six sides; a heptagon has seven sides; an octagon has eight sides; etc.

To sketch polygons, mark off the length of each side at the appropriate angle. Darken the lines to complete the polygon.

## Pictorials

*Pictorial drawings* are three-dimensional representations of objects. They show the three principal measurements of height, length, and depth in one drawing. Three basic types of pictorial drawings used in the building trades are perspective, isometric, and oblique. Common uses of pictorials for residential work include perspective drawings of houses, isometric drawings of plumbing systems, and oblique drawings of cabinets. See Figure 2-3.

**Perspective.** *Perspective drawings* are pictorials with all receding lines converging to vanishing points. These drawings resemble photographs. The number of vanishing points used and location of the object in relation to the horizon determine the type of perspective. Drawings may be made with one, two, or three vanishing points. The object may be located above, on, or below the horizon to produce a worm's-eye, eye-level, or bird's-eye view. Two vanishing points and the eye-level view are commonly used for perspectives of houses. See Figure 2-4.

To sketch perspectives, determine the type to be drawn, and locate the appropriate number of vanishing points. Establish the height of the object, and draw receding lines to converge at the vanishing points. Darken object lines to complete the drawing. The location of the initial height line, its relationship to the horizon, and distance between vanishing points are critical in producing a realistic perspective drawing.

**Isometric.** *Isometric drawings* are pictorials with horizontal lines drawn 30° above (or below) the horizon. Vertical lines remain vertical. All measurements are made on the 30° and vertical axes or lines parallel to them. Non-isometric lines are drawn by locating and connecting their end points on isometric lines.

Circles on isometric drawings appear as ellipses. An *ellipse* is a plane, geometric figure generated by the sum of the distances from two fixed points. Locate these centerpoints to draw ellipses.

To sketch isometrics, draw the isometric axes, and locate end points for the principal measurements of height, length, and depth. Connect these end points to construct the isometric cube. Locate end points for other lines on the isometric axes or lines parallel to them. Connect these end points, and complete all non-isometric lines, circles, and arcs. Darken object lines to complete the drawing. See Figure 2-5.

**Oblique.** *Oblique drawings* are pictorials with one surface drawn in true shape and receding lines projecting back from the face. Receding lines are commonly drawn at 30° or 45° angles. In *oblique cavalier drawings*, receding lines are drawn full-scale. In *oblique cabinet drawings*, receding lines are drawn at one-half scale. Oblique cabinet drawings are the most popular. Circles appear in true shape on the front surface of oblique drawings. They appear as ellipses on receding surfaces.

To sketch oblique cabinets, draw the true shape of the front surface to scale showing the height and length of the object. Draw receding lines to one-half scale. Make all measurements along the oblique axes or lines parallel to them. Draw non-oblique lines by locating their end points on oblique lines. Complete all non-oblique lines, circles, and arcs. Darken object lines to complete the drawing. See Figure 2-6.

## Orthographic Projections

*Orthographic projections* are three-view drawings with each view showing two principal measurements. Each view is a two-dimensional drawing. The front view shows height and length. The side view shows

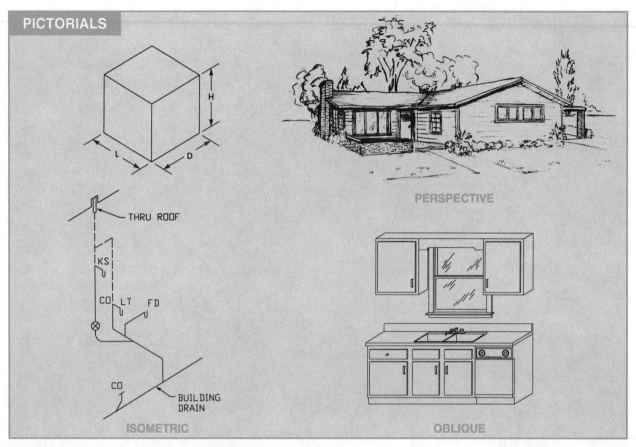

**Figure 2-3.** Pictorial drawings show height, length, and depth.

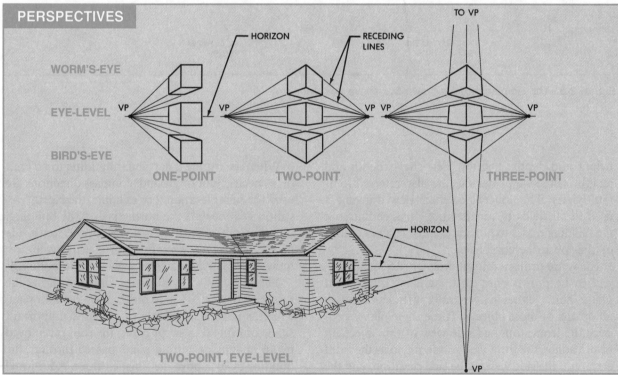

**Figure 2-4.** Receding lines of perspective drawings converge to a vanishing point(s).

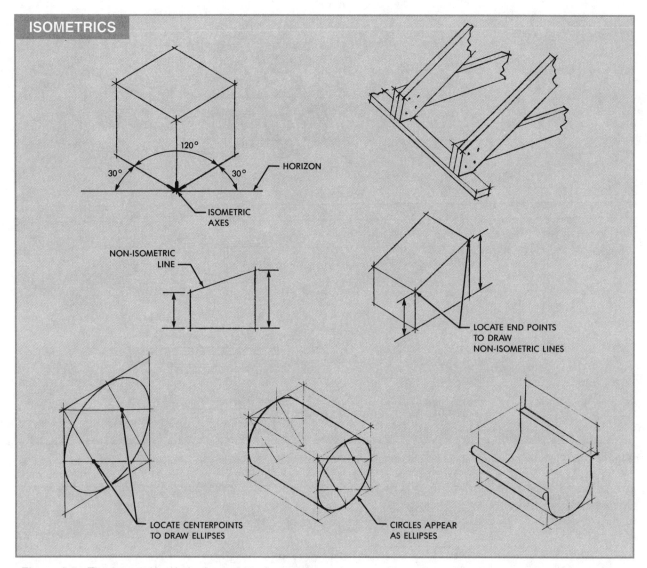

**ISOMETRICS**

**Figure 2-5.** The isometric axis is drawn 30° above horizontal.

height and depth. The top view shows depth and length. Three-view drawings are also referred to as multiviews. The concept of three-view drawing is used in all fields of architecture, engineering, and the building trades to provide graphic representations of the job to be completed.

Lines are projected from every corner of the object to be drawn onto an imaginary, transparent plane. Three planes are generally sufficient to show all details of most objects. These three planes produce the front, top, and side view of the object. In printreading, the front view is referred to as the front elevation, the side view as the side elevation, and the top view as the plan. See Figure 2-7.

When describing a particular building to be built on a specific plot of ground, compass directions are used for each elevation. For example, the North Elevation is a view of the north side of the building, and the South Elevation is a view of the south side of the building. Generally, four elevations are required to show all exterior views of the building.

The top view of a multiview drawing shows the roof plan of a building. In order to see the floor plan, which is generally considered the most important single drawing of a set of plans, the roof is removed by an imaginary cutting plane passed through the building 5'-0" above the finished floor. Additional floor plans for basements and other floor levels of

OBLIQUES

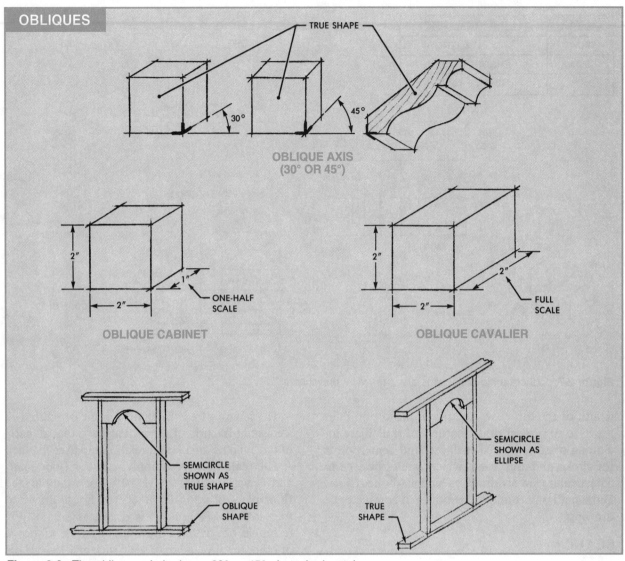

**Figure 2-6.** The oblique axis is drawn 30° or 45° above horizontal.

a multistory house are drawn by using additional imaginary cutting planes.

The elevations and plan views are related by projection lines that connect the parts of one view to another. See Figure 2-8. A simplified isometric box shows the house and indicates the various elevations and plan views. The box is then unfolded to show the relationship of views. Projection lines show that the principal measurements of height, length, and depth are consistent throughout the views.

The relationship of points from one elevation to another aids in the concept of three-view drawings. See Figure 2-9. The simplified pictorial of the tri-level house contains wall surfaces designated A, B,

C, and D. Points on these surfaces are designated with lowercase letters a through h. Surfaces A and B are seen in the front elevation in their true shape and size. Surfaces C and D are seen in the right side elevation in their true shape and size. Points in one view are lines in the other view. For example, the point representing the main ridge is designated ab in the front elevation. This indicates that point a is closer to the observer than point b. The ridge is shown as horizontal line ab in the right side elevation.

To sketch three-view drawings, draw the true shape of the front view to scale showing height and length of the object. Draw projection lines to define the height and depth of the side view and the depth and

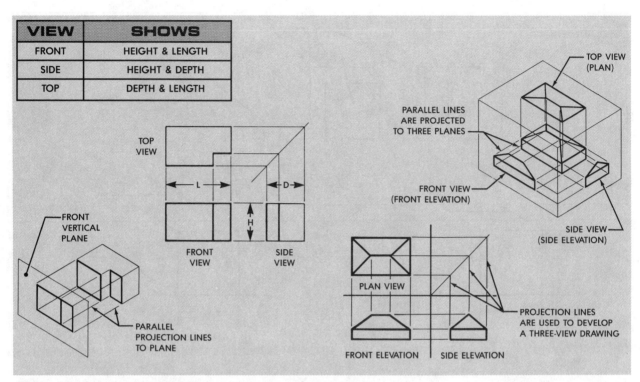

| VIEW | SHOWS |
|------|-------|
| FRONT | HEIGHT & LENGTH |
| SIDE | HEIGHT & DEPTH |
| TOP | DEPTH & LENGTH |

**Figure 2-7.** Orthographic projections are three-view drawings.

length of the top view. Additional projection lines are used to project other features of the object including offsets, slanted surfaces, and centerpoints for circles and arcs. Lines which cannot be seen in a particular view are drawn as hidden (dashed) lines. Darken all object and hidden lines to complete the drawing.

## SCALE

Prints used on the job are reproductions of architectural plans and working drawings drawn to scale. A road map is a common example of a drawing made to scale. An area of several thousand square miles can be shown on a small piece of paper by using a scale of a certain number of miles per inch.

Prints are small enough so that they can be handled easily, yet large enough to show necessary information clearly. Common drawing and print paper sizes in the United States include the following:

| LETTER DESIGNATION | SIZE |
|--------------------|------|
| A | 8½″ × 11″ (sheet) |
| B | 11″ × 17″ (sheet) |
| C | 17″ × 22″ (sheet) |
| D | 22″ × 34″ (sheet) |
| E | 34″ × 44″ (sheet) |
| E+ | 34″ × 44″+ (roll) |

The length of each line on a print is reduced to a constant fraction of its true length so that all parts of the building are in exact relationship to each other.

The scale most commonly used for floor plans and elevations is ¼″ = 1′-0″. For example, the floor plan of a 36′-0″ × 60′-0″ house drawn to the scale of ¼″ = 1′-0″ is drawn as a 9″ × 15″ rectangle. For detailed drawings of doors, windows, or other features, a larger scale such as 1½″ = 1′-0″ is used.

A complete set of plans is seldom drawn to the same scale because of the need to show details at a larger scale or the need to show the roof plan or plot plan at a smaller scale. The scale for a sheet may be shown in the title block and/or below each plan, elevation, detail, or sectional view.

### Architect's Scale

The triangular architect's scale has six ruled faces designed to measure in ten different scales. One of the edges is identical to a 12″ ruler divided into sixteenth's of an inch. The flat architect's scale with beveled edges contains only four ruled faces. This scale is preferred by architects for its convenience, although the triangular architect's scale has more

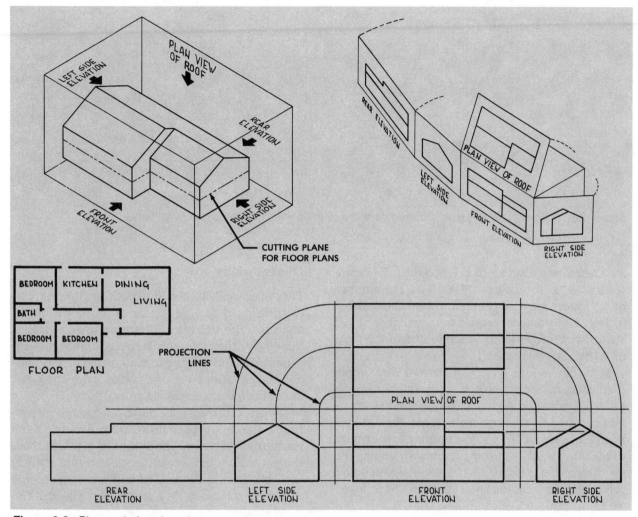

**Figure 2-8.** Plan and elevation views are related by projection.

scales. The ten scales on the triangular architect's scale are:

$$3'' = 1'\text{-}0''$$
$$1\tfrac{1}{2}'' = 1'\text{-}0''$$
$$1'' = 1'\text{-}0''$$
$$\tfrac{3}{4}'' = 1'\text{-}0''$$
$$\tfrac{1}{2}'' = 1'\text{-}0''$$
$$\tfrac{3}{8}'' = 1'\text{-}0''$$
$$\tfrac{1}{4}'' = 1'\text{-}0''$$
$$\tfrac{3}{16}'' = 1'\text{-}0''$$
$$\tfrac{1}{8}'' = 1'\text{-}0''$$
$$\tfrac{3}{32}'' = 1'\text{-}0''$$

Architect's scales are read from left to right and right to left depending on the scale being read. See Figure 2-10. To read the $\tfrac{1}{4}'' = 1'\text{-}0''$ scale, read from right to left beginning at the 0 on the right end of the scale. Notice that the same set of markings is used for the $\tfrac{1}{4}'' = 1'\text{-}0''$ scale and the $\tfrac{1}{8}'' = 1'\text{-}0''$ scale. Read the proper line in relation to the scale

used. For example, $18'\text{-}0''$ on the $\tfrac{1}{4}'' = 1'\text{-}0''$ scale is on the line representing $57'\text{-}0''$ on the $\tfrac{1}{8}'' = 1'\text{-}0''$ scale. Inches are read between the 0 and the end of the architect's scale.

## Tape Measure

A tape measure or any ruler divided into inches and sixteenths of an inch may be used to make and read drawings at the $\tfrac{1}{4}'' = 1'\text{-}0''$ scale, although the results may not be perfectly accurate because the small divisions on the architect's scale are not available. Prints should not be measured to obtain a dimension unless all other methods to obtain the dimension have failed.

Tradesworkers may use a tape measure on the job to work out a specific detail or to transmit information back to the architect. Each $\tfrac{1}{4}''$ space on the tape measure is considered to be $1'\text{-}0''$ on the print. Each $\tfrac{1}{16}''$ space on the tape measure is considered to be

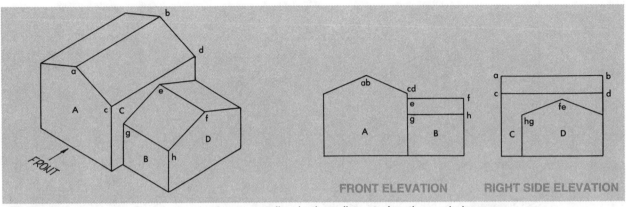

**Figure 2-9.** A point in an elevation is shown as a line in the adjacent elevation and vice versa.

3″. Therefore, a distance of 1 3/16″ on the print represents 4′-9″ at the scale of 1/4″ = 1′-0″. The number of 1/4″ spaces must be counted to find the number of feet. This number is added to the number of 1/16″ spaces which represent 3″ each to obtain the total of 4′-9″. See Figure 2-11.

Some dimensions on a print are found by simple mathematics. For example, if an overall dimension of 26′-0″ is shown for a wall having window openings centered 6′-6″ from each corner, the distance from the center of one window opening to the center of the other window opening is found by adding the given measurements for each window opening and subtracting from the total wall length (6′-6″ + 6′-6″ = 13′-0″; 26′-0″ − 13′-0″ = 13′-0″).

## Symbols and Conventions

Symbols and conventions are drawn to scale to indicate their relative sizes. Walls, windows, doors, plumbing and other fixtures, footings, partitions, chimneys, roofs, and other features are drawn in proportion to their size. See Figure 2-12.

## Dimensioning

The American National Standards Institute (ANSI) has developed specific standards for dimensioning drawings. Through use of these standards, architects can convey their ideas with the assurance that experienced printreaders can read them correctly. Individual variations among drawing styles often lead to confusion when reading prints.

Dimension lines may be terminated by arrowheads, slashes, or dots at extension lines related to points of the drawing. The dimension is placed above the dimension line unless the space is too small. It is then placed in the nearest convenient space and related to the space by a leader. See Figure 2-13.

**Exterior Walls.** Framed, masonry, and masonry veneer walls are dimensioned according to dimensioning standards. The preferred method for framed walls is to dimension to the outside face of stud corner posts. Most of the remaining structural members can be located in relation to these rough framing members. For example, it is easier to locate openings for doors and windows if the dimensions start

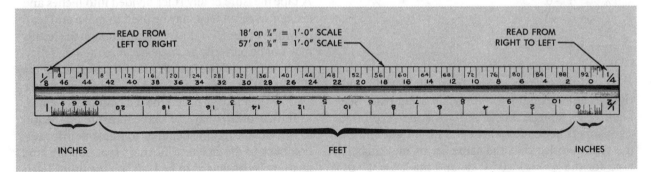

**Figure 2-10.** The architect's scale is used to produce scaled drawings.

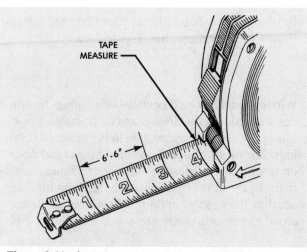

**Figure 2-11.** A tape measure can be used to sketch plans at the ¼″ = 1′-0″ scale.

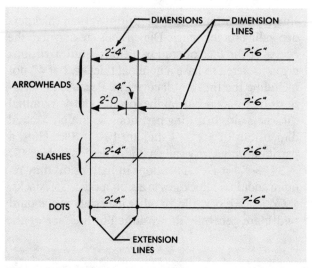

**Figure 2-13.** Arrowheads, slashes, or dots may be used to terminate dimension lines.

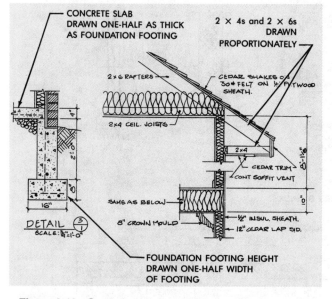

**Figure 2-12.** Symbols and conventions are drawn to scale.

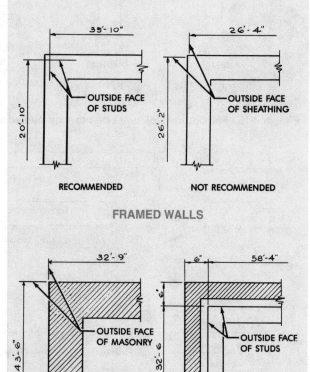

FRAMED WALLS

MASONRY WALLS

**Figure 2-14.** Walls are dimensioned to facilitate construction methods.

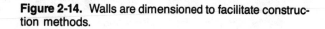

from the outside face of stud corner posts than if the dimensions are given from the outside face of sheathing as the sheathing is applied after the rough openings are located.

Solid masonry walls are dimensioned to their outside faces. Masonry veneer walls are dimensioned to the outside faces of the studs and to the outside faces of the masonry veneer wall. The outside dimensions of the foundation are usually the same as the outside dimensions of the masonry veneer wall. The faces of the studs must be located in relationship to the face of the masonry veneer wall because the frame structure is built before the veneer wall is laid. See Figure 2-14.

**Interior Partitions.** Stud partitions vary in thickness depending on the finish. Dimensions are drawn to the center or the face of stud partitions. Stud partitions are considered to have a nominal thickness of 4″, not including the finish wall covering. Because of variations that may occur on the job site, the 4″ nominal thickness aids in setting partitions accurately. Actual dimensions of a 2 × 4 stud are 1½″ × 3½″. Thus, a partition with ½″ drywall on both sides is 4½″ thick (½″ + 3½″ + ½″ = 4½″). Partitions in bathrooms may require additional thickness to accommodate soil stacks.

Concrete block, tile, or other materials of standard width are generally dimensioned to the face of the finish material. Solid plaster partitions are dimensioned to centerlines. See Figure 2-15.

**Windows and Doors.** Locations of openings for windows and doors on floor plans of frame and brick veneer houses are dimensioned to the center of openings. Locations of openings for windows and doors on floor plans of masonry houses are dimensioned to the center of the openings or to the finish masonry abutting the window or door. Dimensions to the center of the openings are preferred. See Figure 2-16.

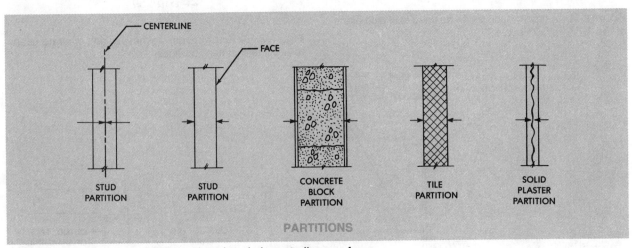

**Figure 2-15.** Partitions are dimensioned to their centerlines or faces.

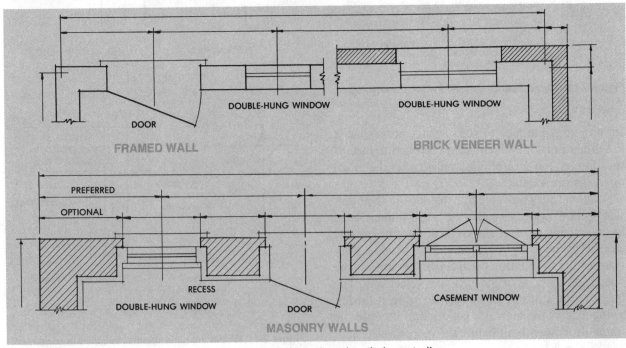

**Figure 2-16.** Openings for windows and doors are dimensioned to their centerlines.

# Sketching

Name _____ Date _____

## Sketching 2-1

**Sketch the missing view of the multiviews.**

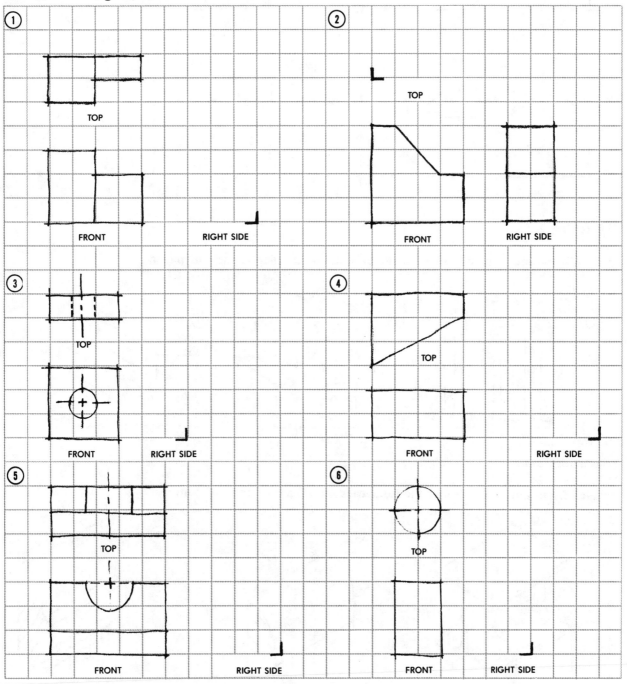

## Sketching 2-2

**Sketch oblique cabinets of the multiviews on a separate sheet of paper.**

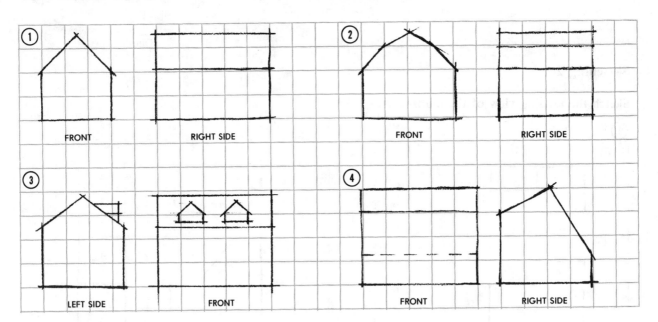

## Sketching 2-3

**Sketch front and right side views of the obliques on a separate sheet of paper.**

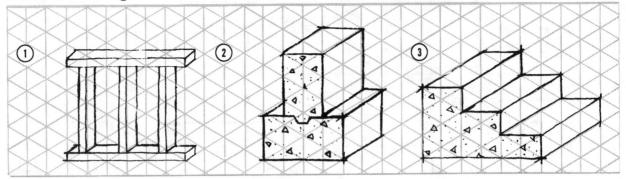

## Sketching 2-4

**Sketch front and right views of the isometrics on a separate sheet of paper.**

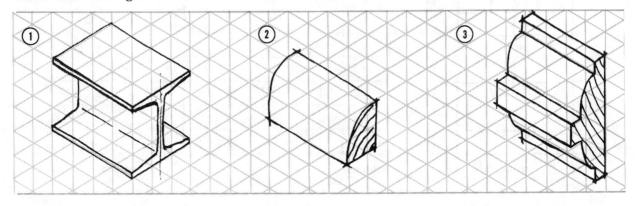

# Sketching 2-5

**Sketch obliques of the sectional views of the moldings. Use 30° receding lines to the right.**

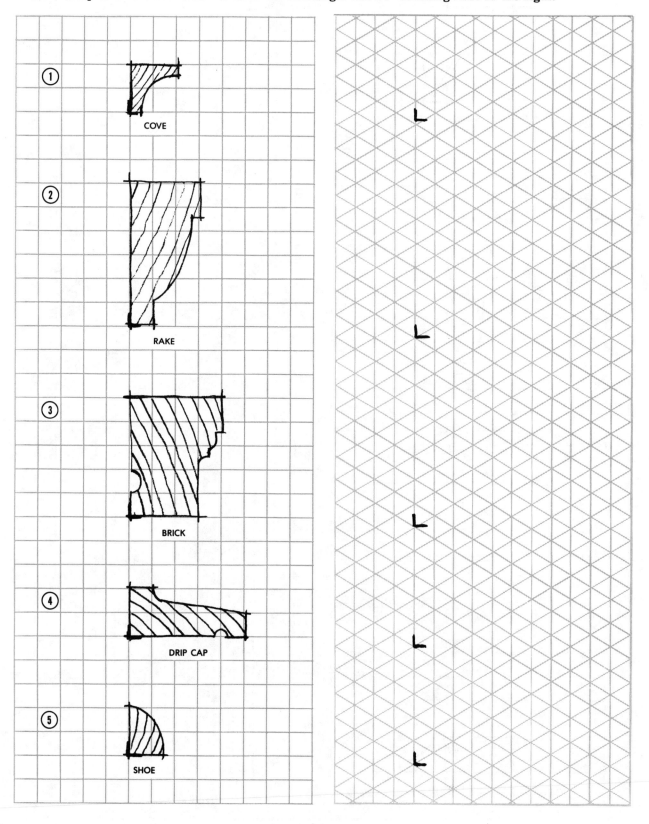

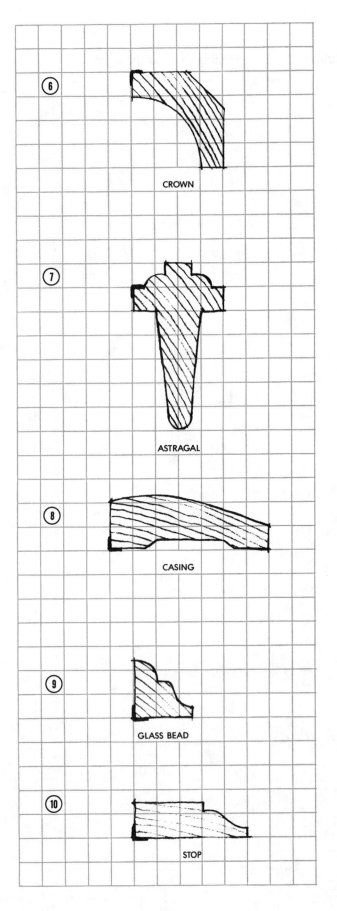

6 CROWN

7 ASTRAGAL

8 CASING

9 GLASS BEAD

10 STOP

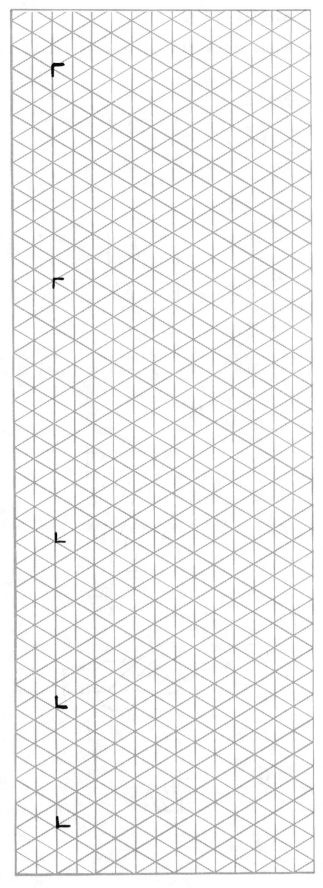

# Review Questions

Name _____ Date _____

## Multiple Choice

_____    **1.** Three basic types of pictorial drawings are _____.
          A. perspective, cabinet, and isometric
          B. isometric, cavalier, and oblique
          C. perspective, isometric, and oblique
          D. isometric, cabinet, and oblique

_____    **2.** C size paper is _____.
          A. 8½″ × 11″
          B. 11″ × 17″
          C. 17″ × 22″
          D. 18″ × 24″

_____    **3.** Isosceles and equilateral triangles are _____ triangles.
          A. acute
          B. obtuse
          C. right
          D. none of the above

_____    **4.** Quadrilaterals are _____ figures.
          A. three-sided plane
          B. three-sided solid
          C. four-sided plane
          D. four-sided solid

_____    **5.** In an oblique cabinet drawing, receding lines are drawn at _____ scale.
          A. one-fourth
          B. one-half
          C. three-fourths
          D. full

_____    **6.** The radius of a circle is _____ the diameter.
          A. the same length as
          B. one-half the length of
          C. twice the length of
          D. two and one-half times the length of

_____    **7.** The top view of a three-view drawing of a house shows the _____ elevation.
          A. front
          B. rear
          C. side
          D. none of the above

8. The preferred method of dimensioning framed walls is to dimension to the _____.
   A. centerlines of stud corner posts
   B. inside face of stud corner posts
   C. outside face of stud corner posts
   D. most convenient location

9. A _____ movement is used when sketching long lines.
   A. smooth wrist
   B. full arm
   C. rigid
   D. none of the above

10. Dimension lines may be terminated by _____.
    A. arrowheads
    B. slashes
    C. dots
    D. all of the above

11. The diameter of a circle is a _____.
    A. straight line passing anywhere through the circle
    B. curved line passing anywhere through the circle
    C. straight line passing through the centerpoint
    D. curved line passing through the centerpoint

12. Each view of a three-view drawing gives _____ principal measurement(s).
    A. one
    B. two
    C. three
    D. four

13. A trapezoid has _____.
    A. opposite sides parallel
    B. opposite side parallel and equal
    C. two sides parallel
    D. two sides parallel and two sides equal

14. A wooden stud partition with ⅜″ drywall on each side of 2 × 4 studs is _____″ thick.
    A. 3½
    B. 4⅛
    C. 4¼
    D. 4½

15. A bird's-eye perspective is drawn _____ the horizon line.
    A. above
    B. on
    C. below
    D. none of the above

## True-False

T   F    **1.** Shading techniques are generally not used when producing three-view drawings.

T   F    **2.** A quadrant is one-half of a circle.

T   F    **3.** An isosceles triangle contains one 90° angle.

T   F    **4.** Compass directions are commonly used to name exterior elevation drawings.

T   F    **5.** An arc is a portion of the circumference of a circle.

T   F    **6.** Oblique drawings may have one, two, or three vanishing points.

T   F    **7.** All circles contain 360°.

T   F    **8.** A complete set of plans is seldom drawn to the same scale.

T   F    **9.** A rhombus contains one 90° angle.

T   F   **10.** Circles on isometric drawings appear as ellipses.

T   F   **11.** Architect's scales are always read from left to right.

T   F   **12.** Solid masonry walls are dimensioned to their outside faces.

T   F   **13.** Rough openings for doors and windows are generally dimensioned to their centerlines.

T   F   **14.** The scale most commonly used for floor plans is ¼″ = 1′-0″.

T   F   **15.** Stud partitions vary in thickness depending on the finish.

## Completion

_____   **1.** _____ drawings include plan, elevation, and sectional views drawn to scale.

_____   **2.** _____ are plane figures with four 90° angles and four equal sides.

_____   **3.** Sketching pencils are either _____ or mechanical.

_____   **4.** Tracing _____ should be used when a sketch is to be duplicated on a diazo printer.

_____   **5.** Semicircles always contain _____°.

_____   **6.** _____ different scales and a ruler are on the triangular architect's scale.

_____   **7.** AutoCAD® is a(n) _____ software program used for drawing.

_____   **8.** _____ lines are drawn level or parallel with the horizon.

_____   **9.** _____ polygons contain sides of equal length.

_____   **10.** Pictorial drawings show the three principal measurements of _____, length, and depth.

_____   **11.** Perspective drawings may have one, two, or three _____ points.

_____   **12.** Concrete block partitions are generally dimensioned to the _____ of the finish material.

_____ 13. The _____ view of a three-view drawing shows the height and depth of an object.

_____ 14. Architects prefer the _____ beveled scale to the triangular scale.

_____ 15. _____ lines are plumb or upright.

_____ 16. Plane figures have _____ surfaces.

_____ 17. Circles on receding surfaces of oblique drawings appear as _____.

_____ 18. Receding lines of oblique drawings may be 30° or _____°.

_____ 19. An acute angle contains less than _____°.

_____ 20. A(n) _____ of a circle is pie-shaped.

_____ 21. _____ are three-sided plane figures.

_____ 22. A straight line always contains _____°.

_____ 23. Right triangles always contain one _____° angle.

_____ 24. _____ lines are straight lines which are neither horizontal nor vertical.

_____ 25. Horizontal lines of isometric drawings are drawn _____° above the horizon.

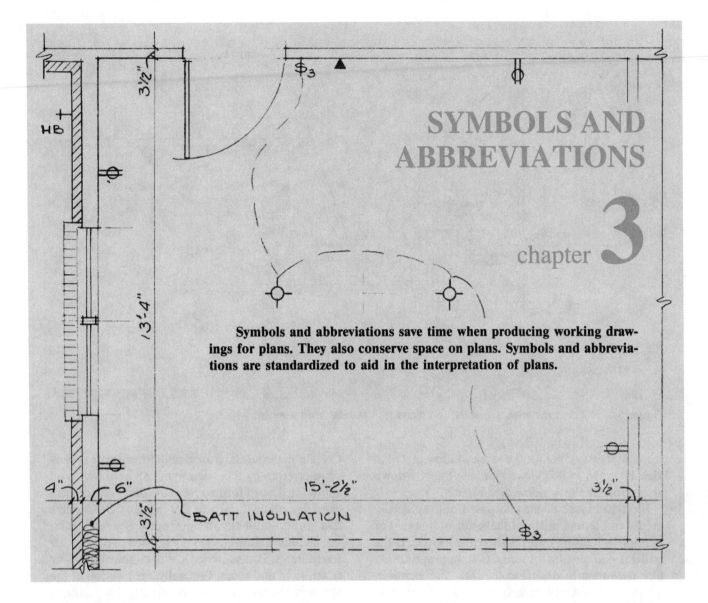

SYMBOLS AND
ABBREVIATIONS

chapter 3

**Symbols and abbreviations save time when producing working drawings for plans. They also conserve space on plans. Symbols and abbreviations are standardized to aid in the interpretation of plans.**

## SYMBOLS

*Symbols* are graphic representations of the components required to complete a building. Symbols conserve space on prints, are easy to draw, and are easily recognized by experienced printreaders. See Appendix. Templates are commercially available for symbols drawn by the conventional method. CAD systems use symbol libraries from which the desired symbol is selected. The American National Standards Institute has standardized symbols in the various trade areas. Materials, equipment, and building parts are shown on prints with symbols. See Figure 3-1.

Symbols for the same material may be drawn differently from plan to elevation to sectional views. For example, the symbol for concrete block is drawn differently for each view. Other materials may have

no standardized symbol for a particular view. For example, there is no standardized symbol for earth in a plan view.

Various symbols are used in combination to show the relationship of building materials. For example, a plan view of a fireplace in a brick veneer wall requires the use of plan symbols for the wood framing and brick to show the brick veneer wall. Additionally, plan symbols showing the face brick, common brick, firebrick, and tile for the hearth are required. See Figure 3-2.

Openings for exterior walls and interior partitions are shown in elevation and plan views. The symbols used to show the doors and windows are drawn differently in these views. For example, the symbol for a double-hung window in an elevation view is drawn as it is viewed looking directly at the surface of the

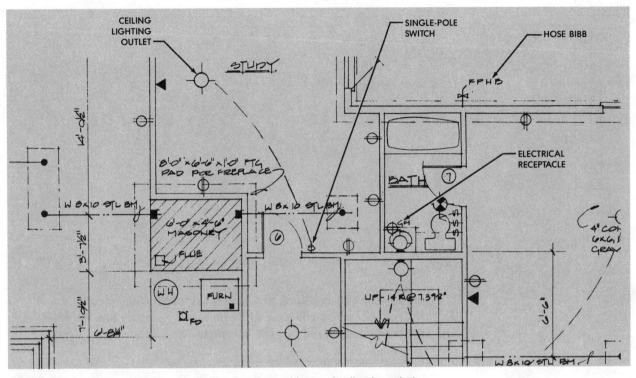

**Figure 3-1.** Building materials are shown on plans with standardized symbols.

window. The symbol for the same window in a plan view is drawn as it is viewed looking directly down at the top of the window. See Figure 3-3.

Electrical symbols are composed of graphic elements and letters of the alphabet or numbers. For example, a lighting outlet is shown as a circle. If the outlet is designed for a specific task, appropriate letters are added to designate the task. For example, a lighting outlet with lampholder is shown with an L. Symbols for electrical switches are shown with letters of the alphabet and numbers. The letter S denotes the switch, and numbers indicate single-pole, double-pole, and so forth. See Appendix.

Standardized symbols for plumbing are graphic elements with letters, as required, to show fixtures. Symbols for piping and valves are highly stylized and simplified line drawings. See Appendix.

## ABBREVIATIONS

*Abbreviations* are key letters of words denoting the complete word. Like symbols, they save time and conserve space on plans. Standardized abbreviations are developed by standards organizations. They are used to denote materials, fixtures, and areas, and to provide simplified instructions to tradesworkers.

Certain materials, fixtures, and other features may be referred to by their acronym. An *acronym* is an abbreviated word formed from the first letter of each word describing the article. For example, GFCI refers to a ground-fault circuit interrupter. See Appendix.

The same abbreviation may be used to denote different items. For example, R is the abbreviation for range, riser, and room. Generally, the location of the abbreviation will indicate its intent. Abbreviations which form a word are followed by a period to avoid confusion. For example, SEW. is the abbreviation for sewer, and KIT. is an abbreviation for kitchen.

Some words have more than one abbreviation. For example, FIN. and FNSH are abbreviations for finish. When two abbreviations are given for the same word, the first abbreviation is preferred. Additionally, some architects may use abbreviations that are not standardized but are fairly obvious based on their location and use.

Symbols and abbreviations are used together to provide comprehensive information in a relatively small space. For example, the shape of a closet is shown in plan view by parallel lines representing the symbol for framed partitions. The door symbol shows the swing of the door, and an abbreviation, either C, CL, or CLOS designates the closet. See Figure 3-4.

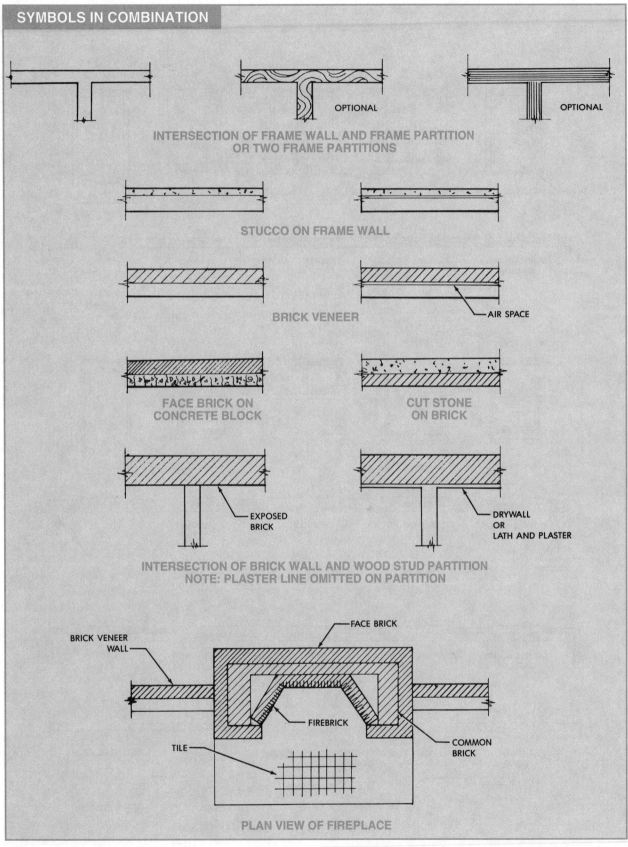

**SYMBOLS IN COMBINATION**

INTERSECTION OF FRAME WALL AND FRAME PARTITION
OR TWO FRAME PARTITIONS

OPTIONAL

OPTIONAL

STUCCO ON FRAME WALL

BRICK VENEER

AIR SPACE

FACE BRICK ON
CONCRETE BLOCK

CUT STONE
ON BRICK

EXPOSED
BRICK

DRYWALL
OR
LATH AND PLASTER

INTERSECTION OF BRICK WALL AND WOOD STUD PARTITION
NOTE: PLASTER LINE OMITTED ON PARTITION

FACE BRICK

BRICK VENEER
WALL

FIREBRICK

COMMON
BRICK

TILE

PLAN VIEW OF FIREPLACE

**Figure 3-2.** Symbols may be used in combination.

## EXTERIOR WALLS AND INTERIOR PARTITIONS

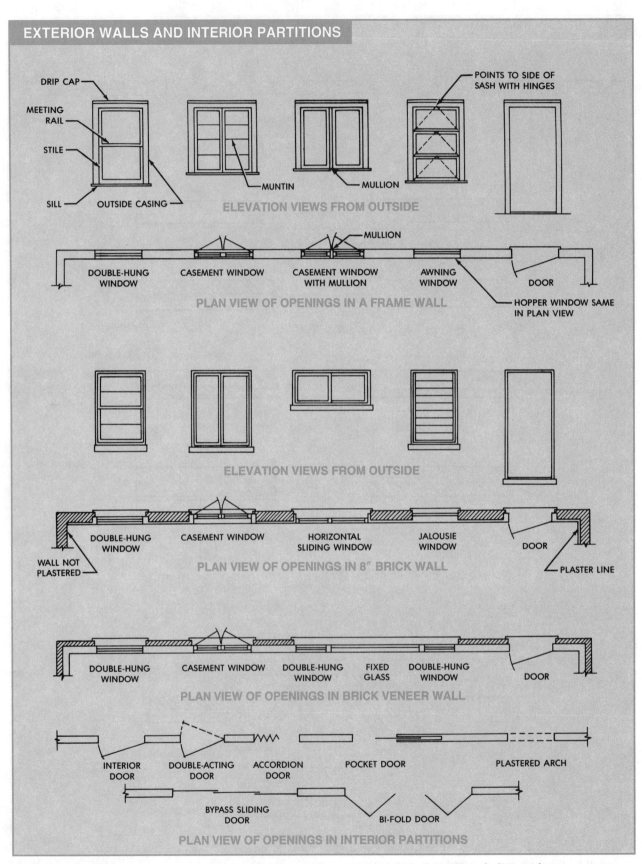

**Figure 3-3.** Different symbols are used in elevation and floor plans to show the same window or door.

# SYMBOLS AND ABBREVIATIONS

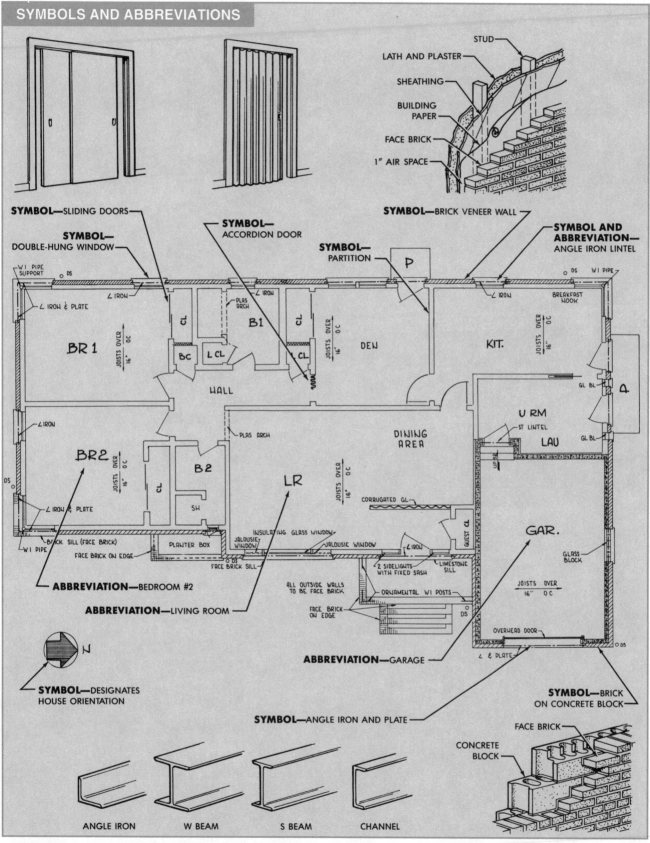

**Figure 3-4.** Symbols and abbreviations are used together to provide comprehensive information.

# Sketching

Name _____ Date _____

## Sketching 3-1

**Complete the chart by sketching the missing symbols. Refer to the Appendix.**

| | ELEVATION | PLAN | SECTION |
|---|---|---|---|
| **CONCRETE** | | ① | SAME AS PLAN VIEW |
| **CONCRETE BLOCK** | ② | | ③ |
| **WOOD** | SIDING  PANEL ④ | PARTITION ⑤ | OR<br>ROUGH MEMBER ⑥  TRIM MEMBER ⑦ |
| **STRUCTURAL CLAY TILE** | ⑧ | | SAME AS PLAN VIEW |
| **GLASS** | | ⑨ | SMALL SCALE  LARGE SCALE ⑩ |

51

## Sketching 3-2

**Sketch the symbol indicated. Refer to the Appendix.**

**For example:** Three-way switch        $S_3$

1. Finish grade

2. Bush

3. Fence

4. Natural grade

5. Point of beginning

6. Property line

7. Water heater

8. Hose bibb

9. Lighting outlet (ceiling)

10. Junction box

11. Range outlet

12. Single-pole switch

13. Duplex receptacle outlet

14. Lighting outlet (wall)

15. Motor

16. Exposed radiator

17. Supply duct

18. Return duct

19. Thermostat

20. Steam

# Review Questions

Name _____ Date _____

## Identification

**Refer to the Appendix.**

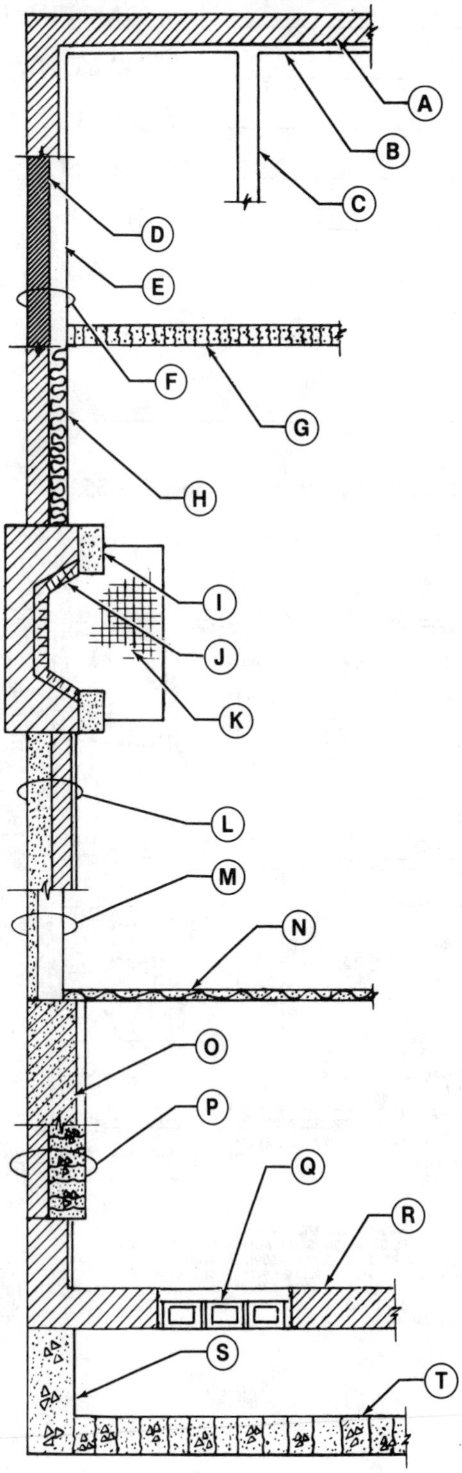

_____ **1.** Brick wall

_____ **2.** Cut stone

_____ **3.** Common brick

_____ **4.** Brick veneer wall

_____ **5.** Concrete block

_____ **6.** Insulation

_____ **7.** Drywall

_____ **8.** Firebrick

_____ **9.** Concrete

_____ **10.** Wood stud partition

_____ **11.** Rubble stone

_____ **12.** Face brick

_____ **13.** Stucco on frame wall

_____ **14.** Common brick on concrete block wall

_____ **15.** Wood stud wall

_____ **16.** Tile

_____ **17.** Glass block

_____ **18.** Cut stone on common brick wall

_____ **19.** Solid plaster partition

_____ **20.** Tile partition

## Abbreviations 3-1

**Write the word(s) for the abbreviations only in the following notations.**

| | | | |
|---|---|---|---|
| _____ | **1.** AL SASH | _____ | **9.** COMMON BRK |
| _____ | **2.** 16" C TO C | _____ | **10.** 14 R UP |
| _____ | **3.** OAK FIN. FLR | _____ | **11.** $\frac{5}{8}$" SC |
| _____ | **4.** WALL-HUNG WC | _____ | **12.** REF CAB. |
| _____ | **5.** $\frac{1}{2}$" CI | _____ | **13.** PRCST CONC |
| _____ | **6.** 2400 SQ FT | _____ | **14.** T & G SIDING |
| _____ | **7.** BR 1 | _____ | **15.** 4 × 10 BM |
| _____ | **8.** $\frac{1}{2}$" AB | | |

**Refer to Printreading on page 55.**

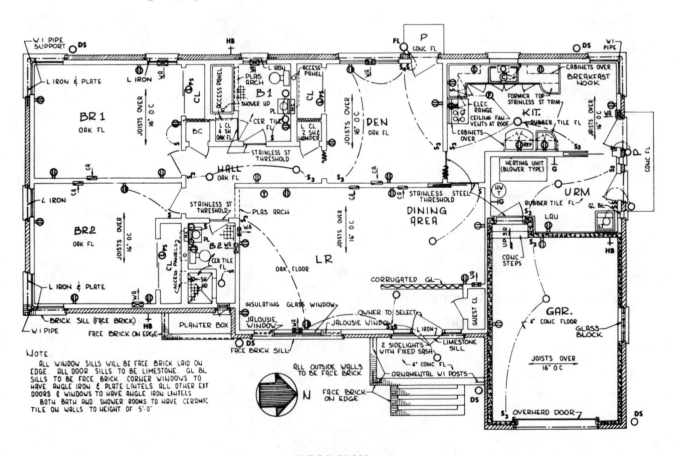

**FLOOR PLAN
SMITH RESIDENCE**

# PRINTREADING

**Refer to Smith Residence—Floor Plan on page 54.**

## Completion 3-1

**Complete the chart to list the number of outlets, fixtures, and switches shown in each room.**

| | ○ | ⊖ | (L)_PS | ⊖_R | ⊖ | (F) | (C) | ○_FL | (T) | S | S₃ | S₄ |
|---|---|---|---|---|---|---|---|---|---|---|---|---|
| *Example:* Living Room—Dining Area | 2 | 6 | | | 1 | | | | 1 | 2 | 2 | |
| 1. Bathroom 2 | | | | | | | | | | | | |
| 2. Bedroom 2 | | | | | | | | | | | | |
| 3. Bedroom 1 Closet | | | | | | | | | | | | |
| 4. Bedroom 2 | | | | | | | | | | | | |
| 5. Bedroom 2 Closet | | | | | | | | | | | | |
| 6. Hall | | | | | | | | | | | | |
| 7. Bathroom 1 | | | | | | | | | | | | |
| 8. Den | | | | | | | | | | | | |
| 9. Den Closet | | | | | | | | | | | | |
| 10. Kitchen | | | | | | | | | | | | |
| 11. Utility Room | | | | | | | | | | | | |
| 12. Garage | | | | | | | | | | | | |
| 13. Exterior Lights | | | | | | | | | | | | |
| 14. Total | | | | | | | | | | | | |

*Note:* Include example in total.

## True-False

| | | | |
|---|---|---|---|
| T | F | **1.** | The hot water tank is located in the kitchen. |
| T | F | **2.** | Soil stacks are located in partitions behind toilets. |
| T | F | **3.** | Four hose bibbs are shown. |
| T | F | **4.** | LT indicates a light table in the utility room. |
| T | F | **5.** | Medicine cabinets are located over lavatories in the bathrooms. |
| T | F | **6.** | The house is heated by a forced warm air system. |
| T | F | **7.** | Nine warm air registers are shown on the floor plan. |
| T | F | **8.** | Nine cold air registers are shown on the floor plan. |
| T | F | **9.** | The electric range is built into the kitchen cabinets. |
| T | F | **10.** | A refrigerator cabinet is shown over the refrigerator. |

T    F    **11.** Seven stainless steel thresholds are shown on the floor plan.

T    F    **12.** Rubber tile flooring is used in the kitchen and utility room.

T    F    **13.** Ceramic tile flooring is used in the bathrooms.

T    F    **14.** Ceiling joists are spaced 16″ OC.

T    F    **15.** A planter box is shown on the East wall.

## Multiple Choice

**1.** The _____ have windows facing East.
A. U RM, BR 1, and LR
B. LR, KIT., and U RM
C. BR 1, BR 2, and DEN
D. LR, BR 2, and B 2

**2.** Not counting the U RM, HALL, B 1, B 2, and GAR., the house has _____ rooms.
A. three
B. four
C. five
D. six

**3.** The shortest route from BR 1 to the front door is _____.
A. HALL, DEN, LR
B. HALL, LR
C. HALL, BR 2, LR
D. none of the above

**4.** Exterior walls, excluding the garage, are _____.
A. concrete block
B. brick on concrete block
C. brick veneer
D. siding

**5.** _____ walls are shown for the garage.
A. Framed
B. Brick veneer
C. Concrete block
D. Brick on concrete block

**6.** A _____ partition is shown between the L CL and the DEN CL.
A. brick
B. paneled
C. drywall
D. solid plaster

**7.** Lintels over windows are _____.
A. angle iron, and angle iron and plate
B. angle iron only
C. plate only
D. none of the above

_____    **8.** Ceiling joists over the LR run _____.
      A. North and South
      B. East and West
      C. North and South and East and West
      D. ceiling joist direction is not indicated

_____    **9.** Ceiling joists over the GAR. _____.
      A. run East and West
      B. are spaced 16″ OC
      C. are not indicated on the floor plan
      D. none of the above

_____    **10.** Not including the GAR. door, _____ exterior doors are shown on the floor plan.
      A. two
      B. three
      C. four
      D. five

_____    **11.** Glass block is located in the GAR. and the _____.
      A. DEN
      B. KIT.
      C. U RM
      D. DINING AREA

_____    **12.** Windows with fixed sash are located _____.
      A. on both sides of the DEN door
      B. on both sides of the front door
      C. both A and B
      D. neither A nor B

_____    **13.** B 1 has a _____ window.
      A. hopper
      B. casement
      C. awning
      D. double-hung

_____    **14.** An accordion door is shown _____.
      A. between the HALL and BR 1
      B. between the HALL and B 2
      C. for each L CL in the HALL
      D. between the DEN and KIT.

_____    **15.** Hose bibbs will be shown on the _____ elevation view(s).
      A. East
      B. East and West
      C. North
      D. North and South

_____    **16.** Regarding the GAR., _____.
      A. a casement window is shown on the North wall
      B. three risers lead to the U RM door
      C. the 6″ concrete floor is brush-finished
      D. none of the above

_____  17. Regarding the windows, _____.
　　　　A. corner windows have angle iron and plate lintels
　　　　B. three corner windows are shown
　　　　C. both A and B
　　　　D. neither A nor B

_____  18. Regarding the plumbing, _____.
　　　　A. each bathroom contains a tub-shower combination
　　　　B. B2 is adjacent to the DEN
　　　　C. access panels are located in closets
　　　　D. all of the above

_____  19. Regarding the scale, _____.
　　　　A. the Smith Residence is drawn to the scale of ⅛″ = 1′- 0″
　　　　B. the Smith Residence is drawn to the scale of ¼″ = 1′- 0″
　　　　C. no scale is shown
　　　　D. none of the above

_____  20. Regarding the masonry, _____.
　　　　A. a face brick sill is shown beneath the LR windows
　　　　B. all outside walls are face brick
　　　　C. face brick on edge finishes the planter box
　　　　D. all of the above

## Completion 3-2

_____  1. The hot water tank is located in the _____ room.

_____  2. The front of the Smith Residence faces _____.

_____  3. Tub and shower walls are finished with ceramic tile to a height of _____.

_____  4. A(n) _____ sill is shown at the front door.

_____  5. The L CL near the DEN has two shelves and one _____.

_____  6. _____ floors are shown in all bedrooms.

_____  7. Glass block sills are _____ brick.

_____  8. Three ornamental _____ posts are noted on the front porch.

_____  9. A(n) _____ door separates the KIT. and U RM.

_____  10. The Smith Residence is a(n) _____-story dwelling.

## Abbreviations 3-2

**Write the abbreviated word(s) only in the space provided.**

_____  1. COMMON BRK WALL　　_____  6. GL BLOCK

_____  2. 14 R DN　　_____  7. BLDG LOT 12

_____  3. CONC FLR　　_____  8. STUCCO FIN.

_____  4. CLG LINE　　_____  9. 3½″ TYP

_____  5. EXT　　_____  10. RGH OPNG

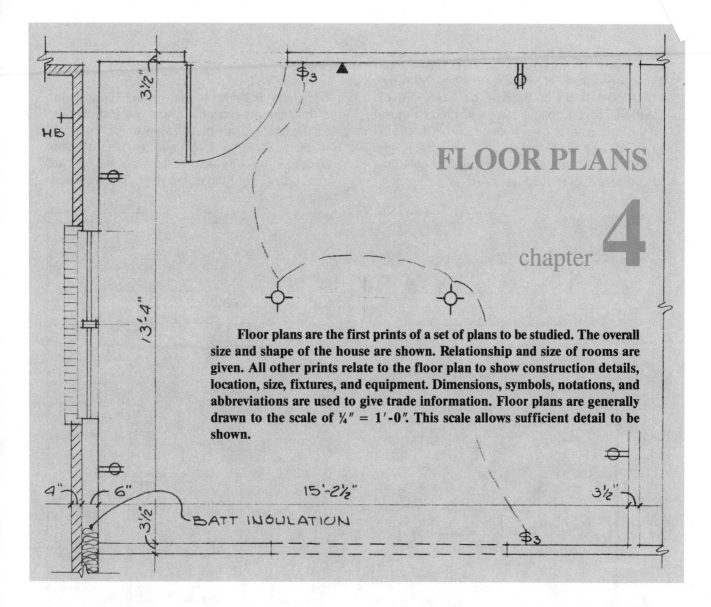

# FLOOR PLANS

chapter 4

Floor plans are the first prints of a set of plans to be studied. The overall size and shape of the house are shown. Relationship and size of rooms are given. All other prints relate to the floor plan to show construction details, location, size, fixtures, and equipment. Dimensions, symbols, notations, and abbreviations are used to give trade information. Floor plans are generally drawn to the scale of ¼″ = 1′-0″. This scale allows sufficient detail to be shown.

## FLOOR PLANS

*Floor plans* are scaled views of the various floors in a house looking directly down on the house as if the roof were removed. They are of primary importance as they show the broad aspects of shape, size and relationship of rooms, and the layout of auxiliary space such as hallways, stairs, and closets. The layout of attached garages, carports, patios, and decks is also shown. Floor plans are generally the first set of plans to be drawn and the most used set on the construction site.

Dimensions on floor plans give the overall size of the house and location of all wall offsets, partitions, doors, and windows. Wall and partition thicknesses are dimensioned. Room sizes are shown. The location and plan sizes of closets, fireplaces, stairs, and other features are also dimensioned.

Symbols on floor plans represent material, equipment, and fixtures. Floor tile, glass block, and common or face brick are examples of materials that are shown with symbols. Electrical equipment, such as receptacles and outlets, and plumbing equipment, such as hose bibbs and floor drains, are shown with symbols. Fixtures such as kitchen sinks, bathroom lavatories, tubs, showers, and water closets are also shown with symbols.

Abbreviations on floor plans conserve space and help prevent cluttering. LR (living room), BR (bedroom), DR (dining room), and B (bath) are common abbreviations used to designate rooms. Additionally, abbreviations are used to designate material, equipment, and fixtures.

Notations (notes) on floor plans may be general, typical, or specific. A *general note* such as ALL

DIMENSIONS TO FACE OF STUDS refers to all dimensioned studs. A *typical note* such as PROVIDE FURRED SOFFIT OVER ALL KITCHEN WALL CABINETS refers to all kitchen wall cabinets shown on the floor plan. A *specific note* such as CER TILE FLR (ceramic tile floor) refers to that floor only. Note that abbreviations may be used in notations to conserve space. See Figure 4-1.

## Scale

Floor plans are drawn to exact scale. The scale for the floor plan is generally given as a general note near the name of the plan. For example, the notation SCALE: ¼″ = 1′-0″ may be placed below FLOOR PLAN. The name and scale of the plan are placed near the bottom of the sheet. The scale may

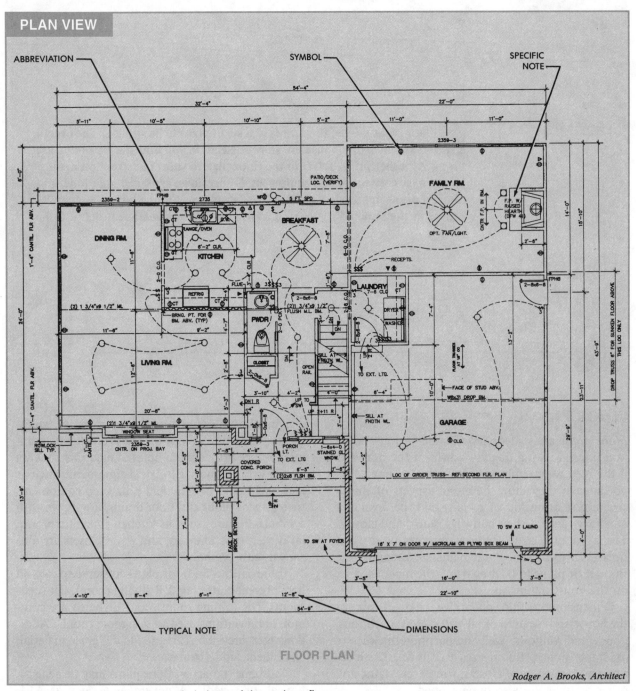

*Rodger A. Brooks, Architect*

**Figure 4-1.** Floor plans are scaled views of the various floors.

also be given in the title block. All rooms, hallways, stairs, and other areas are drawn to the same scale so that they are in correct size relationship to one another.

The most commonly used scale for floor plans is ¼″ = 1′-0″ (1/48th scale). This size scale yields a floor plan large enough to convey sufficient information on standard size print sheets. Dimensions on the scaled drawing give the actual size of the rooms, hallways, and so forth. For larger houses, scales of ⅛″ = 1′-0″ (1/96th scale) or 3⁄16″ = 1′-0″ (1/72nd scale) may be used.

## Cutting Plane

Floor plan views are horizontal sections made by an imaginary cutting plane taken through the house 5′-0″ above the finished floor. The cutting plane passes through walls, partitions, the upper sash of windows, kitchen wall cabinets, medicine chests in bathrooms, and so forth. Solid lines show features below the cutting plane. Dashed, or hidden, lines show features above the cutting plane. For example, kitchen base cabinets are shown with solid lines while kitchen wall cabinets are shown with dashed lines. See Figure 4-2.

A significant departure from the idea of looking down at a slice taken through the house to see the floor plan is followed in noting structural members. For example, the note 2 × 10 JOISTS OVER on a first floor plan indicates that the joists are overhead and support the second floor.

## Orientation

Floor plans are generally drawn so that the front view of the house is toward the bottom of the sheet. This follows the basic concept of orthographic drawing. An orthographic drawing contains at least two, and more commonly, three views. Floor plans may be oriented to fit the print sheet. For example, when a house is long and narrow, such as one to be built on a narrow city lot, the front is usually placed facing toward the right edge of the print sheet.

## Relationship

Floor plans are related to each other. Structural provisions are made so that the load of floors and partitions is transferred to supporting members or partitions immediately below. Stairs are designed so that they start on one floor and end in the proper place on the floor above or below. Heating ducts are designed to start at the furnace, pass through first floor partitions, and end at registers in the desired locations in second floor partitions or floors.

Floor plans and exterior elevations are drawn to the same scale and are exactly related to each other.

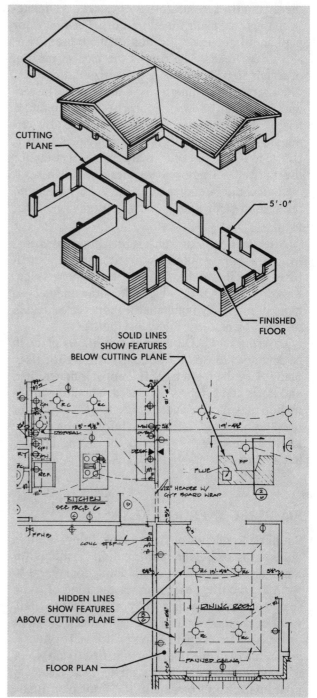

**Figure 4-2.** The cutting plane is 5′-0″ above the finished floor.

Windows that appear on the floor plan are the same size and the same distance from the building corners as they are on the elevations. The house is visualized by referring back and forth from floor plans to elevations.

The same information is generally not shown in two places. For example, a stairway that rises from the first to the second story is not shown completely on the floor plans for either story. The first floor plan shows the exact location of the bottom riser and a few treads. The *riser* is the vertical portion of a stair step. The *tread* is the horizontal portion. The stairway is terminated on a break line. The second floor plan shows the top riser location and a few of the descending stair treads which also terminate on a break line. A notation, such as 16 R UP or 16 R DN gives the total number of risers for the stairway. A window that appears on a stair landing and could be shown on the first floor plan *and* the second floor plan is shown on only one floor plan.

After the first floor plan has been drawn, it is used to locate walls, windows, stairways, and other features to facilitate drawing the second floor plan. Layering on CAD drawings allows the architect to quickly and easily maintain the proper relationship of floor plans for a multistory house.

Layering on CAD drawings also allows plans to be developed for specific trade areas from basic floor plans. For example, a basic floor plan is drawn first, and notes and symbols are added. Plumbing information may then be added to create a floor plan for the plumbing trades. Electrical information may be added to the basic floor plan for the electrical trades. See Figure 4-3.

## SIMPLIFIED FLOOR PLANS

Simplified sketches of floor plans do not contain dimensions or detailed information. Symbols for materials of construction, windows, doors, and so forth are modified to permit visualization of room shapes and relationships without distracting detail. A set of prints should be studied in much the same order as a person might inspect a house. For houses with more than one floor plan, the first (main) floor plan is studied first. For example, in a one-story house with a full basement, the floor plan (main floor plan) is studied first, and the basement (foundation) plan is studied next.

Pictorial sketches are useful in determining room shapes and relationships when studying floor plans. Such sketches show basic shapes and relationships but little detail. Overall concepts can be developed from the sketches.

### One-Story House

A one-story house has one floor plan. Two floor plans are required if the house has a basement. The floor plan is drawn as if a cutting plane had been passed through the house 5'-0" above the finished floor and the upper part removed so that the printreader could look directly down on the floor. Sketches may be completed by the conventional or CAD method. See Figure 4-4.

Entry to the house is through either the living room door or kitchen door. The main entry is the living room door. A coat closet (CL) to the right of the living room door serves as a windbreak and divides the entry area from the dining area. Picture windows are placed along the front of the house.

The dining area is between the living room and the kitchen. It is not enclosed by partitions. One window is shown in the dining area wall. The kitchen contains L-shaped base and wall cabinets. Wall cabinets are shown by dashed lines as they are above the cutting plane. Light and ventilation are provided by a window on the wall opposite the cabinets.

Three bedrooms and a bath are located off the hall. The hall is entered from the living room. The entry to the hall is located so that it is not directly across from the bathroom door or the door to Bedroom 1. A linen closet is located at the end of the hall near the bathroom. This space could be devoted to a furnace if required. (Note that a chimney for the furnace is shown on the roof in the pictorial sketch.)

All bedrooms are well-lighted with large windows. Bedrooms 2 and 3 have cross ventilation. Closets for Bedrooms 1 and 2 are located between the two bedrooms. The closet for Bedroom 3 is larger. Bi-fold or sliding doors are required for this closet. All bathroom fixtures have been arranged along the common partition with the kitchen. The backed-up plumbing reduces material and labor costs.

### Two-Story House

A two-story house with a basement has three floor plans. See Figure 4-5. The three cutting planes which

## LAYERING OF CAD DRAWINGS

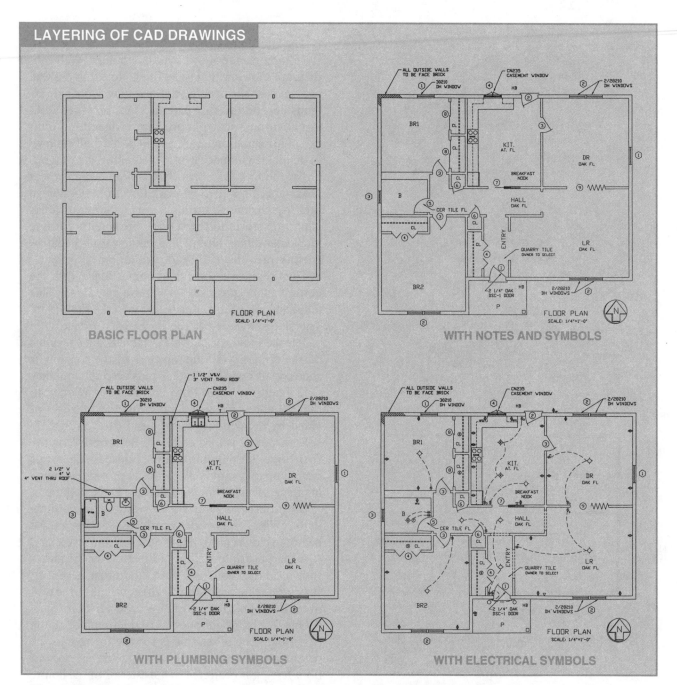

**Figure 4-3.** Layering of CAD drawings allows plans to be easily developed for specific trades.

produce the basement, first floor, and second floor plans are indicated by cutting plane lines designated *A-A, B-B,* and *C-C,* respectively. *A-A* is the projection for the basement plan. It passes through the basement windows. *B-B* is the projection for the first floor plan. It passes through the house at the upper sash of windows and upper part of doors on the main floor. *C-C* is the projection for the second floor

plan. It passes through the house at the upper sash of windows on the upper floor.

The first floor plan shows that the living and dining rooms are one large L-shaped room. The kitchen, bathroom, and rear exit occupy the remaining portion of the first floor. Stairs to the second floor begin from a platform one riser above the floor level. The stairway is open with a railing on the living room

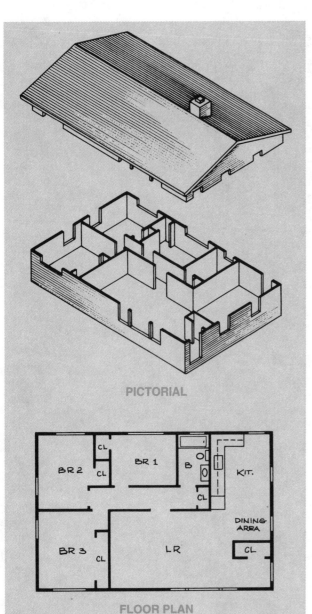

PICTORIAL

FLOOR PLAN

| BR2 | CL | BR 1 | B | KIT. |
| CL | | | | |
| | | CL | DINING AREA |
| BR 3 | CL | LR | CL |

**Figure 4-4.** A one-story house has one floor plan.

The kitchen has U-shaped base and wall cabinets. Base cabinets are shown with solid lines. Wall cabinets are shown with dashed lines. Note that the wall cabinet near the opening to the dining room side extends beyond the base cabinet. This provides additional storage space above the refrigerator. Additional information about cabinets is usually shown in detail views. Space is provided along the wall opposite the sink for a breakfast table. The appliances are arranged for efficient work flow from the storage area (refrigerator) through the preparation areas (sink and stove) to the serving areas (either in the kitchen or dining room). The window over the kitchen sink provides light and ventilation.

The fireplace in the living room is drawn as if the cutting plane passes 1'-0" above the floor. This allows details of its shape to show on the floor plan. The flue shown is for the basement furnace.

The second floor plan shows the layout of rooms on the upper level. The top riser of the stairway is shown in its exact location. Treads descend until they stop against a break line. The stairway is open to the hall and is protected by a railing. Light is provided by the window at the head of the stairs.

Bathroom 2 is located above Bathroom 1 on the first floor. All plumbing in the house is stacked or backed up to common partitions to provide economy in piping. A small window is located in the bathroom.

Bedrooms 1, 2, and 3 have windows in both exterior walls. This provides good cross ventilation. All bedroom closets have sliding doors to conserve space.

A load-bearing partition, directly over the girder shown on the first floor plan, and other necessary structural members (not shown) continue to the wall over the stairs to support overhead joists.

One flue in the chimney is for the fireplace on the first floor. The other flue is for the furnace in the basement. Before the chimney passes the second floor level, it is projected to the exterior of the house.

The basement floor plan shows information about the foundation, windows, stairs, and building structure. The location of the bottom riser of the stairway is shown. Some of the treads are drawn. They terminate against a break line.

The steel beam directly below the load-bearing partition on the first floor supports the joists which run 90° to the beam. This beam does not continue to the wall at the stairs as it would pass through the stairway. The notation 2 × 10 FLOOR JOISTS OVER and the directional symbol give the size and direction of floor joists.

side. Stairs to the basement descend from the rear hallway and are directly below the stairs to the second floor. A platform is provided as a safety measure inside the doorway before the stairs begin. Both sets of stairs stop against break lines.

The construction of the building requires support for second floor joists across the center of the house. A *joist* is a framing member that supports a floor. A load-bearing partition and a built-up girder support the floor joists. A *built-up girder* is made of laminated wooden boards designed to carry heavy loads.

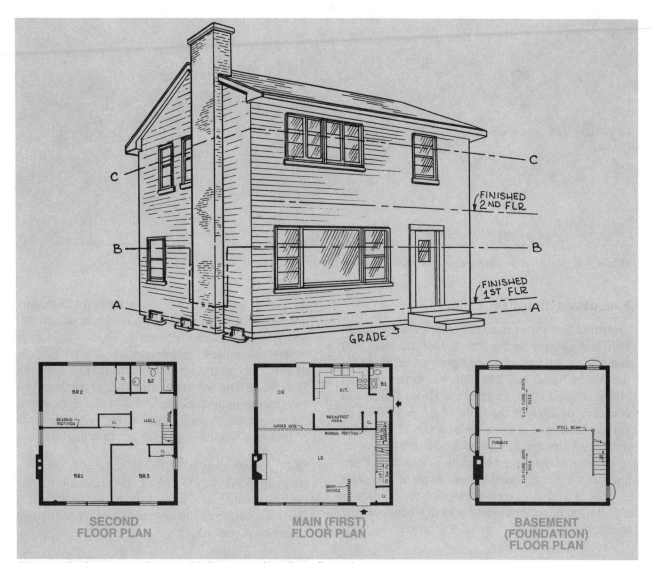

**Figure 4-5.** A two-story house with basement has three floor plans.

The fireplace foundation is carried down to the footing with only the furnace flue showing. A *footing* is a support base for a foundation wall. Basement windows are located in areaways to provide light and ventilation.

### One-and-One-Half Story House

A one-floor house with a steeply sloped roof provides useful attic space which may be converted into living space. For example, a traditional Cape Cod house uses attic space for living space. Dormers are added to provide additional floor area, ventilation, and architectural effect. *Dormers* are projections from sloping roofs that provide additional interior area. They are described by their roof types. Three common types of dormers are the gable-end, hipped-end, and shed. See Figure 4-6.

Two cutting planes are required for a Cape Cod house. Each cutting plane is taken 5'-0" above its finished floor. The cutting plane for the second floor plan reveals the irregular outline of dormered walls. See Figure 4-7. The floor plans are then read in the same way as other floor plans.

### READING FLOOR PLANS

The Wayne Residence is a contemporary house with a full basement. Refer to Wayne Residence, Sheets 1 and 2. Sheet 1 shows the Foundation/Basement Plan. Sheet 2 shows the Floor Plan.

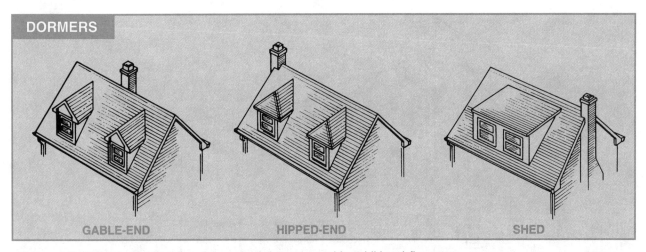

**Figure 4-6.** Gable-end, hipped-end, and shed dormers provide additional floor area.

## Foundation/Basement Plan

The title block shows that the Foundation/Basement Plan is drawing number 1 of 7. This indicates that seven sheets comprise the complete set of plans for the Wayne Residence. The plan was drawn by Pam Hulen (initials PLH) of Hulen & Hulen Designs in February of 1989 at the scale of $\frac{1}{4}'' = 1'-0''$. The Wayne Residence is to be built on Lot 12 of Country Club Fairways in Columbia, Missouri.

The title of the sheet, Foundation/Basement Plan, and the scale is repeated below the plan. The scale is $\frac{1}{4}'' = 1'-0''$. An arrow showing North indicates orientation of the house and is used when referring to exterior elevation views of the house.

**Foundation.** The foundation wall varies in thickness around the house while the footings remain constant in size. See Basement Wall and Walkout Details, Sheet 5. Brick veneer is shown on the South and East sides of the foundation wall. A stepped 6″ brick ledge provides a base for the brick veneer. The West foundation wall is below grade level. Retaining walls are shown on the East and South walls. These walls compensate for the drop in grade due to the slope of the lot. See Elevations, Sheets 3 and 4.

The North wall is the rear wall. See North Elevation, Sheet 4. A lot that has an 8′-0″ drop in grade from front to rear gives builders an opportunity to provide a walkout basement. A *walkout basement* has standard size windows to provide light and standard size doors for entry and exit. Such a basement can be finished to provide additional living space at a competitive square foot cost. The lot slope and use of retaining walls allows a walkout basement in

the Wayne Residence. A window well provides clearance for an awning window in the future workshop. See Window Schedule, Sheet 5.

Steel columns, 4″ in diameter, are placed in foundation footings to carry the beams that support the floor joists. The W beams (wide flange) are described by their nominal measurement over the flanges and the weight per running foot. A W 8 × 10 beam is 8″ high and weighs 10 pounds per running foot.

**Slab.** A typical note, 4″ CONC SLAB W/ 6 × 6, 10 × 10 WWF ON GRAVEL BASE, is shown for the basement floor. This note is repeated for the unfinished portion of the basement and for the garage slab. See Appendix.

**Stairway.** Entry to the basement from the main floor of the house is by the stairs. A 2′-6″ × 6′-8″ × 1⅜″ hollow-core, six panel, wood door is at the foot of the stairs. Panels in these doors are formed from moldable fiberboard with wood components used in hollow-core construction. Natural or various wood finishes are available. The stairway contains 14 risers at 7.392″ each. A ceiling outlet above the landing is controlled by three-way switches.

**Family Room.** A 4′-9″ × 4′-0½″ vinyl-clad, casement window and a 6′-0″ × 6′-8″ aluminum-clad sliding glass door provide natural light to this room. Electrical receptacles are located in all walls. The ceiling outlet is shown with its switch leg. A telephone outlet, represented by a triangle, is shown on the West wall.

**Bath.** A 2′-4″ × 6′-8″ × 1⅜″ hollow-core, six panel, wood door leads from the family room to the bath. The bath contains a water closet, vanity with lavatory, and tub. A GFI receptacle is located near the lavatory, and wall switches control a combination ceiling light and fan.

**Study.** A 2'-6" × 6'-8" × 1⅜" hollow-core, six panel, wood door leads to the study. Casement windows, marked B, provide light and ventilation. Electrical receptacles are located in framed walls. The ceiling outlet is controlled by a switch located near the door. Future plans of the owner include boxing-in the column and steel beam and adding a floor-to-ceiling

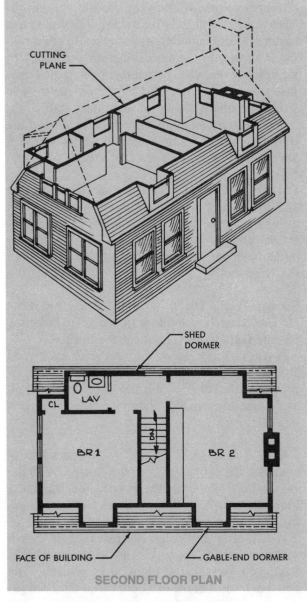

**Figure 4-7.** Cutting planes on the upper floor of dormered-roof houses reveal the irregular wall line.

combination storage and bookcase unit in the recessed area on the East wall. Alternate plans call for making a fourth bedroom in this room.

**Unfinished Basement.** The storage area and future workshop are left unfinished to reduce the initial cost of the house. The water heater, furnace, and fireplace foundation with flue are completed. The power panel is located on the rear wall. A floor drain is located near the water heater.

Registers for supply and return air are not shown on this set of plans. The mechanical subcontractor will design and install the heating and cooling system. The contractor will coordinate the mechanical subcontractor's work with other trades to assure that proper electrical and structural requirements are met.

## Floor Plan

Title block information given on the Foundation/Basement Plan is repeated in the title block. The Floor Plan is drawn to the scale of ¼" = 1'-0". The arrow shows that the house faces South.

Overall dimensions of the house are 44'-4" × 72'-8". Dimensions are shown for locating all doors and windows. These dimensions are given to the centers of the openings. A 24'-0" × 12'-0" deck is located on the North side of the house. Doors from Bedroom 1 and the living room open to the deck.

The main floor of the Wayne Residence contains the following:

| | |
|---|---|
| 1. Entry | 8. Living Room |
| 2. Hallway | 9. Dining Room |
| 3. Bedroom 1 | 10. Kitchen |
| 4. Bedroom 2 | 11. Breakfast Area |
| 5. Bedroom 3 | 12. Laundry |
| 6. Bath | 13. Screened Porch |
| 7. Master Bath | 14. Garage |

**Entry and Hallway.** The house is entered from a covered stoop through a 3'-0" × 6'-8" × 1¾" exterior door. The door is metal-insulated and includes a deadbolt. Light is provided to the entry by 15" sidelights. Wall switches control exterior and interior ceiling outlets and the ceiling outlet in the coat closet. A rod and shelf are shown in the coat closet.

The hallway provides access to the basement, bath, and bedrooms. Dashed lines show a 30" × 30" scuttle in the ceiling. A *scuttle* is an opening in a ceiling or roof with a removable or movable cover. Dashed lines in the linen closet indicate shelf rods. Lighting for the hallway is provided by two ceiling outlets controlled by three-way switches.

**Bedrooms.** Three 2'-6" × 6'-8" × 1⅜" hollow-core, six panel wood doors lead to the bedrooms. Bedroom 1 is the largest bedroom. It measures 11'-9½" × 17'-0". Bedrooms 2 and 3 are 11'-2" × 12'-3½". Electrical receptacles, ceiling outlets with switch legs, and telephone outlets are shown in each bedroom. The closet in Bedroom 2 has bi-fold doors. Bedrooms 1 and 3 have walk-in closets. A small linen closet is also shown for Bedroom 1.

**Bathrooms.** The master bath contains a water closet, tub, and shower separated from the double-bowl lavatory. Wall cabinets, mirror, and soffit are shown in Elevations 2-5 and 3-5. Ceiling outlets in the master bath are controlled by wall switches. A combination ceiling light and fan is also controlled by a wall switch. GFI receptacles are placed near the lavatories.

The hall bath contains a lavatory, water closet, and tub and shower combination. The combination ceiling light and fan is controlled by a wall switch. A GFI receptacle is placed near the lavatory. See Elevation 1-5.

**Living Room.** The 19'-4½" × 21'-6" living room is located at the rear of the house. Four aluminum-framed patio replacement windows allow light to enter while providing a view of the backyard. The deck, which is adjacent to the living room, is reached by a full glass door.

A masonry fireplace defines the South portion of the living room. Cutting planes refer to Fireplace Details 1-6 and 2-6, which contain additional information.

Recessed ceiling outlets are controlled by three-way switches located on the East wall near the stair landing and near the patio door. Electrical receptacles are spaced around walls and wired so as not to interfere with the location of a future door to the breakfast area. A header is provided for the future door. Dashed lines show a gypsum board wrap on the header over the opening leading to the kitchen. A telephone outlet is shown on the West wall.

**Dining Room.** The panned ceiling with recessed lighting provides a formal look for this room. A *panned ceiling* contains two ceiling levels connected by sloped surfaces. Refer to the Panned Ceiling De-

tail, Sheet 6. Vinyl-clad casement windows provide light and ventilation.

**Kitchen and Breakfast Area.** The kitchen is arranged with L-shaped, straight-run, and island base cabinets. Wall cabinets are indicated by dashed lines. Refer to Sheet 6. The arrangement of storage, preparation, and serving areas is compact and efficient. A desk on the East wall provides planning space. The telephone outlet is located in the wall above the desktop.

GFI receptacles are located adjacent to the sink. The disposal in the sink is controlled by a wall switch. A range receptacle serves the drop-in stove in the island cabinet. The stove has a downdraft vent system to remove cooking odors and smoke. Recessed ceiling fixtures over the bar cabinet are controlled by a single-pole switch. The ceiling outlet in the breakfast area is also controlled by a three-way switch. A clock outlet is shown on the South wall.

The laundry, screened porch, and garage are entered from the kitchen. Refer to the Door Schedule, Sheet 5 for types and sizes of doors shown.

**Laundry.** The laundry contains space for the washer and dryer with wall cabinets above. A 240 V receptacle supplies the dryer. The laundry includes a pantry for additional storage. A pull switch lighting outlet is shown. A casement window is centered in the West wall.

**Screened Porch.** The 10'-4" × 14'-0" screened porch has a 4" concrete slab that is sloped from the house. See the Foundation/Basement Plan, Sheet 1.

A screened door leads to the backyard, and a wood atrium door leads into the breakfast area. A weatherproof electrical receptacle is shown on the East wall. The casement window above the kitchen sink is vinyl-clad.

**Garage.** The 23'-4" × 21'-2½" garage has a 4" concrete slab that slopes 2" from the North wall to the garage door. Refer to the Foundation/Basement Plan, Sheet 1. A steel beam lintel provides a base for brick veneer above the overhead door.

A frostproof hose bibb is located on the North wall and electrical receptacles are located on North, East, and West walls. Lighting is controlled by switches located at the two doors. The overhead garage door is operated by a garage door opener to be connected to the switched receptacle in the ceiling.

# Sketching

Name _____ Date _____

**Sketch the following on Warner Residence—Floor Plan. For example, the ceiling outlet (13) in the dining room is controlled by two three-way switches (15).**

1. Brick exterior walls
   *Note: All exterior walls are brick.*
2. Double-hung window
3. Casement window
4. Exterior door
5. Interior door
6. Pocket door
7. Accordion door
8. Bypass sliding door

9. Bi-fold doors
10. Cased opening
11. Water closet
12. Telephone
13. Ceiling outlet
14. Recessed ceiling outlet
15. Three-way switch
16. 120 V receptacle
17. 240 V receptacle

18. 120 V GFI receptacle
19. Pull-chain switch
20. WP receptacle
21. Ceiling exhaust
22. Kitchen sink
23. Bathtub
24. Hose bibb
25. Shelf and rod

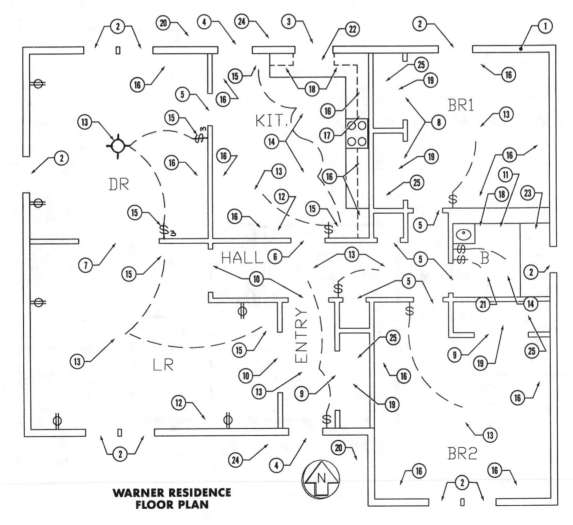

**WARNER RESIDENCE
FLOOR PLAN**

**Refer to Printreading 4-1 on page 76.**

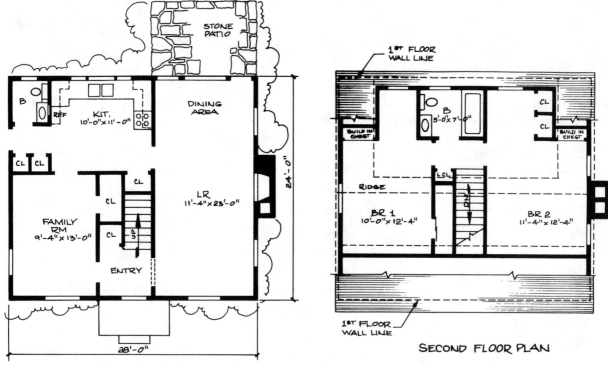

FIRST FLOOR PLAN

SECOND FLOOR PLAN

**TORRANCE RESIDENCE**

**Refer to Printreading 4-2 on page 77.**

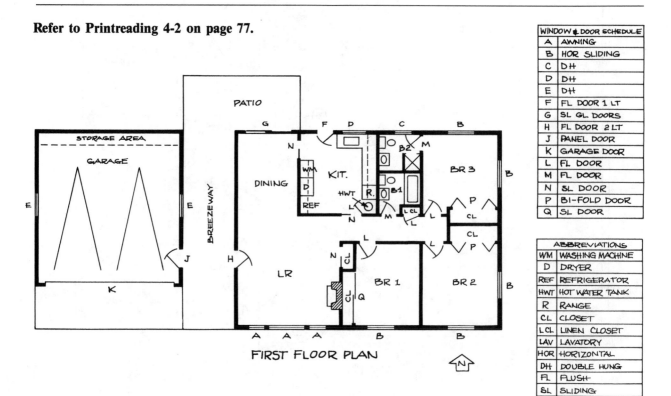

FIRST FLOOR PLAN

**CHAPMAN RESIDENCE**

| WINDOW & DOOR SCHEDULE | |
|---|---|
| A | AWNING |
| B | HOR SLIDING |
| C | DH |
| D | DH |
| E | DH |
| F | FL DOOR 1 LT |
| G | SL GL DOORS |
| H | FL DOOR 2 LT |
| J | PANEL DOOR |
| K | GARAGE DOOR |
| L | FL DOOR |
| M | FL DOOR |
| N | SL DOOR |
| P | BI-FOLD DOOR |
| Q | SL DOOR |

| ABBREVIATIONS | |
|---|---|
| WM | WASHING MACHINE |
| D | DRYER |
| REF | REFRIGERATOR |
| HWT | HOT WATER TANK |
| R | RANGE |
| CL | CLOSET |
| LCL | LINEN CLOSET |
| LAV | LAVATORY |
| HOR | HORIZONTAL |
| DH | DOUBLE HUNG |
| FL | FLUSH |
| SL | SLIDING |
| GL | GLASS |

# Review Questions

Name _____ Date _____

## True-False

T    F    **1.** A portion of the stairway from the first to the second floor is shown on both floor plans.

T    F    **2.** Exterior elevations are generally drawn to a smaller scale than floor plans.

T    F    **3.** The broad aspects of shape, size, and relationship of rooms are shown on floor plans.

T    F    **4.** Layering on CAD drawings refers to the total area of the house.

T    F    **5.** Electrical fixtures and outlets on floor plans are shown with abbreviations only.

T    F    **6.** Cutting planes for floor plans are taken $3'-0''$ above the finished floor.

T    F    **7.** Floor plans may be oriented to fit the print sheet.

T    F    **8.** Stairways in plan views are terminated on a break line.

T    F    **9.** The note $2 \times 10$ JOISTS OVER on a first floor plan indicates that the joists are overhead and support the second floor.

T    F    **10.** Floor plans are generally drawn so that the front view of the house is toward the bottom of the sheet.

T    F    **11.** Abbreviations are not used on floor plans.

T    F    **12.** The cutting plane for second floor plans passes through the house at the upper sash of windows on the main floor.

T    F    **13.** A two-story house with basement requires three floor plans.

T    F    **14.** Sketches for floor plans may be completed by the conventional or CAD method.

T    F    **15.** The tread is the vertical portion of a stair step.

T    F    **16.** Floor plans are generally drawn to the scale of $\frac{1}{4}'' = 1'-0''$.

T    F    **17.** A typical note gives information pertaining to all such items on the floor plan.

T    F    **18.** Dashed lines on floor plans show features above the cutting plane.

T    F    **19.** The scale for each room of a floor plan varies to fit the size of the room.

T    F    **20.** The first floor plan is generally studied before second floor or basement floor plans.

## PRINTREADING 4-1

**Refer to Torrance Residence on page 74.**

_____     1. A _____ house is shown.
          A. one-story
          B. one-story with basement
          C. one-and-one-half story
          D. two-story with basement

_____     2. The living room measures _____ × _____.

_____
_____     3. The arch between the entry and the living room is shown with _____ lines.

_____     4. The upstairs bathroom contains a _____.
          A. shower, tub, and water closet
          B. vanity, tub, and water closet
          C. vanity, shower, and water closet
          D. vanity and water closet only

T    F     5. Part of the ceiling in Bedrooms 1 and 2 is sloped.

T    F     6. A dormer is shown on the rear of the house.

T    F     7. A dormer is shown on the front of the house.

T    F     8. The face of the rear dormer is directly above the face of the first floor wall line.

_____     9. The house is _____ wide across the front.

_____     10. The patio is laid with _____.
          A. brick
          B. stone
          C. stone with brick details
          D. brick with stone details

_____     11. Bedroom _____ is the largest bedroom.

_____     12. Two built-in _____ are shown on the upper floor.

T    F     13. Two risers are shown from the grade to the front door.

T    F     14. Stairs to the upper floor begin in the living room.

T    F     15. Four exterior doorways are shown on the first floor plan.

_____     16. The dining area is _____.
          A. part of the living room
          B. adjacent to the patio
          C. adjacent to the kitchen
          D. all of the above

_____     17. The family room has _____.
          A. two windows on the front wall
          B. two windows on the side wall
          C. one window on each wall
          D. no windows

_____ **18.** The house measures _____ from front to back.

T  F  **19.** The kitchen sink is centered on the kitchen windows.

T  F  **20.** The upstairs bathroom measures $5'\text{-}6'' \times 7'\text{-}0''$.

## PRINTREADING 4-2

**Refer to Chapman Residence on page 74.**

_____ **1.** The house has _____.
  A. a full basement
  B. two fireplaces
  C. a two-car garage
  D. two levels

_____ **2.** The living room _____.
  A. is separated from the dining room by a room divider
  B. has direct access to Bedroom 1
  C. has a fireplace on the West wall
  D. is part of a living-dining room area

_____ **3.** The dining area has _____.
  A. one awning window
  B. a swinging door to the kitchen
  C. sliding glass doors to the patio
  D. two casement windows

_____ **4.** The kitchen _____.
  A. has wall cabinets above the sink
  B. has two flush exterior doors
  C. is U-shaped
  D. also serves as a laundry

_____ **5.** The hallway _____.
  A. opens to all bedrooms
  B. has no door that swings out into it
  C. is L-shaped
  D. has two linen closets

_____ **6.** Regarding closets, _____.
  A. Bedroom 1 has two closets
  B. linen closets have sliding doors
  C. Bedrooms 2 and 3 have closets of equal size
  D. Bathroom 2 has one linen closet

_____ **7.** Regarding plumbing, _____.
  A. all fixtures in the bathrooms are located along the same wall
  B. two tubs are shown
  C. the kitchen sink is located under the kitchen window
  D. the washing machine is located on the East wall of the kitchen

_____ 8. The South Elevation has _____ windows.
   A. double-hung
   B. hopper
   C. casement
   D. awning and horizontal sliding

_____ 9. The East Elevation has _____ windows.
   A. awning
   B. horizontal sliding
   C. double-hung and fixed sash
   D. double-hung and casement

_____ 10. The North Elevation has _____ windows.
   A. horizontal sliding and double-hung
   B. horizontal sliding and casement
   C. double-hung and casement
   D. awning and double-hung

## Identification

**Refer to the Appendix.**

_____ 1. Ceiling outlet

_____ 2. Bi-fold doors

_____ 3. Double-hung window

_____ 4. 240 V receptacle

_____ 5. Ceiling exhaust and light

_____ 6. Common brick

_____ 7. Floor drain

_____ 8. Weatherproof receptacle

_____ 9. 120 V receptacle

_____ 10. Double-acting door

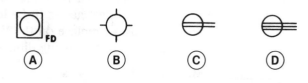

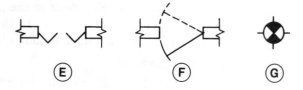

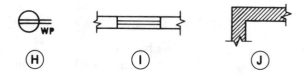

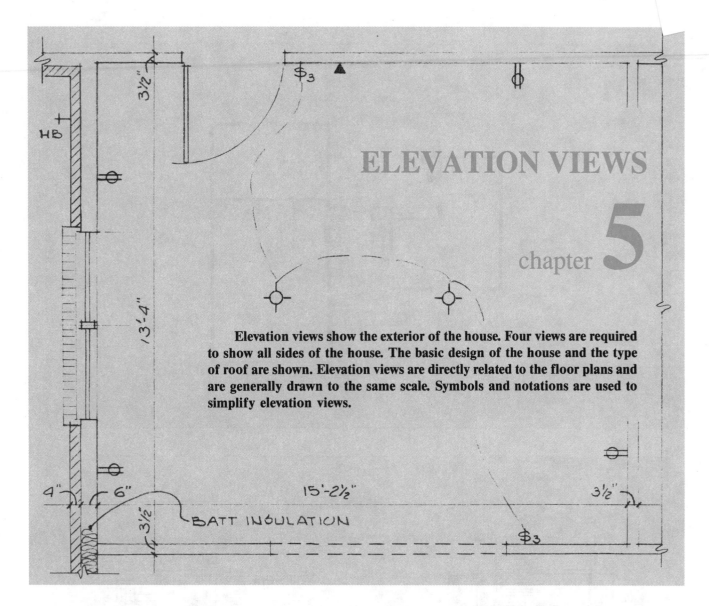

**ELEVATION VIEWS**

chapter **5**

Elevation views show the exterior of the house. Four views are required to show all sides of the house. The basic design of the house and the type of roof are shown. Elevation views are directly related to the floor plans and are generally drawn to the same scale. Symbols and notations are used to simplify elevation views.

## ELEVATIONS

*Elevations* (or elevation views) are orthographic projections showing the exterior view of one side of a building. Elevations may also be made of an interior wall or feature of a building. The four exterior elevations, each showing a side of the building, are part of the working drawings prepared by the architect. They show what the building will look like when it is completed. Their function is to show the basic design of the house, where openings are placed, and the materials used.

North, East, South, and West are the four major points on a compass. These are used to designate elevations. The North Elevation is the elevation facing North, not the direction a person faces to see that side of the house. The East Elevation is the elevation facing East, the South Elevation is the eleva-

tion facing South, and the West Elevation is the elevation facing West. See Figure 5-1.

The various prints of a set of working drawings are related to each other. It is often necessary to refer to several prints in order to find all the information on one subject. Original drawings are drawn so that the foundation plan, floor plans, and elevations exactly match regarding the location of windows, doors, and other details. Information found on elevation views includes the following:

Design of the Building

General shape. Location of offsets, ells, patios, decks, steps, porches, bays, dormers, chimneys, and other features.

Dimensions and materials for footings and foundations.

**PLAN VIEW**

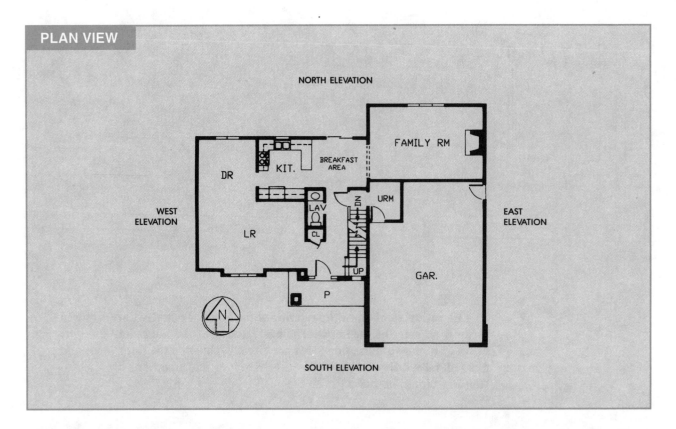

**ELEVATION VIEWS**

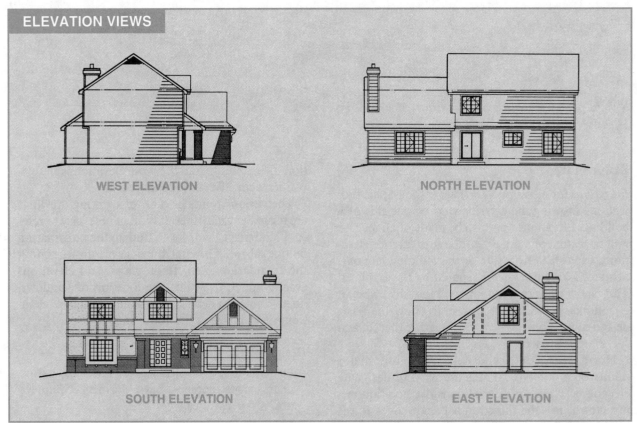

**Figure 5-1.** Elevation views show the exterior sides of a building.

### Roof

Type, slope, material, vents, gravel stops, projection of eaves, and so forth.

### Openings

Windows: type, size, swing, and location.

Doors: type, size, and location.

### Dimensions

From established grade to finished basement floor level.

From established grade to finished first floor level.

Floor-to-floor heights.

Heights of special windows above floor.

From ridge to top of chimney.

### Exterior Finish—Material

Type of siding: concrete, concrete block, brick, stone, stucco, or other material.

### Exterior Finish—Trim

Treatment of windows, entrance doorways, columns, posts, balustrades, cornices, and other features.

### Miscellaneous Details

Electrical fixtures, utility outlets, hose bibbs, gutters and downspouts, flashing and waterproofing.

### Symbols

Symbols showing building materials and fixtures save space on elevation views. Most elevation views for residential construction are drawn to the scale of ¼″ = 1′-0″. This scale is 1/48th the size of the actual building. Such a scale dictates the use of symbols to clearly show the materials used. See Figure 5-2. Refer to the Appendix.

### Abbreviations

Abbreviations are also used to conserve space on drawings. They may be used alone, in notations, or with symbols to describe materials. Standardized abbreviations are used to avoid misinterpretation. Only uppercase letters are used in abbreviations. Abbreviations that form a word are followed by a period. For example, the abbreviation for inch is IN. See the Appendix.

## BUILDING DESIGN

Houses are generally designed from the inside out. The floor plan is developed first based upon the living space required. After the room arrangement has been determined, the exterior of the house is planned. Local ordinances regarding lot size, setback, and

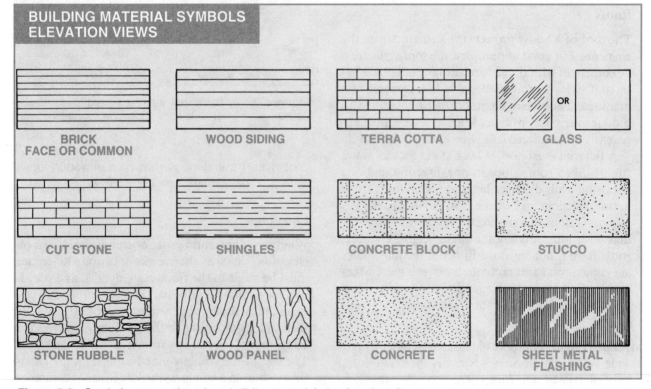

**BUILDING MATERIAL SYMBOLS ELEVATION VIEWS**

BRICK FACE OR COMMON  WOOD SIDING  TERRA COTTA  GLASS

CUT STONE  SHINGLES  CONCRETE BLOCK  STUCCO

STONE RUBBLE  WOOD PANEL  CONCRETE  SHEET METAL FLASHING

**Figure 5-2.** Symbols are used to show building materials in elevation views.

house size are considered. The preference of the client, type of materials to be used, and expense are reviewed.

### Styles

Houses are often designed in a particular style. For example, houses may be traditional or contemporary. A *traditional house* reflects long-standing design elements. For example, Cape Cod, Colonial, and Old English designs remain popular today because of their design elements. A *contemporary house* reflects current trends in design. The ranch house and other new designs making extensive use of glass and solar heating are examples of contemporary design.

Details such as the general proportions, type of roof, windows, and trim must be consistent with the design. Elevation views show the elements of style and design. Whenever a change is shown in a wall or roof line, a modification of the basic rectangular shape of the house is indicated.

All visible lines in elevation views are drawn as solid lines. Any part of the building below grade level is shown with dashed lines. For example, foundation footings and walls, and areaways for windows below grade are shown with dashed lines.

### Roofs

The roof of a house protects the structure from the elements. For good appearance, the roof style must be consistent with the style of the house. The six basic roof styles are flat, shed, gable, hip, gambrel, and mansard. These may be modified or combined to produce various roof lines. Roof styles are most apparent on elevation views. See Figure 5-3.

A flat roof must slope at least ⅛" per foot for water runoff. Shed roofs slope in one direction, making opposing walls different heights. Gable roofs slope in two directions and are the most popular roof style. Hip roofs slope in four directions. All walls of buildings with hip roofs are the same height. Gambrel roofs have a double slope in two directions. They are popular on barns and country-style houses. Mansard roofs have a double slope in four directions. They are used for multistory dwellings.

**Slope (Pitch).** All roofs must slope in order to provide water runoff. *Slope (pitch)* is the relationship of roof rise to run. *Rise* is the vertical increase in height. *Run* is the horizontal distance in which the

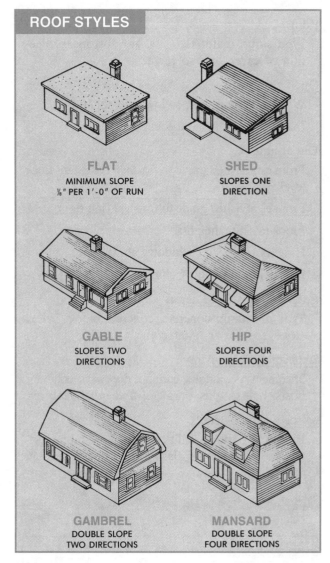

ROOF STYLES

FLAT
MINIMUM SLOPE
⅛" PER 1'-0" OF RUN

SHED
SLOPES ONE
DIRECTION

GABLE
SLOPES TWO
DIRECTIONS

HIP
SLOPES FOUR
DIRECTIONS

GAMBREL
DOUBLE SLOPE
TWO DIRECTIONS

MANSARD
DOUBLE SLOPE
FOUR DIRECTIONS

**Figure 5-3.** Six basic roof styles are used in residential construction. Roof styles are shown on elevation views.

roof rises. Roof slope is shown on elevation views with a slope symbol. See Figure 5-4.

### Openings

Windows, doors, and other openings are shown on elevation views in their exact location. Reference must be made to the floor plan, details, and schedules for additional information. For example, an elevation view shows what a window looks like and its vertical location in the wall. The floor plan gives the dimension of the window from a corner of the house. The size of the window may be given on the elevation view or in a window schedule which is commonly located on another sheet. See Figure 5-5.

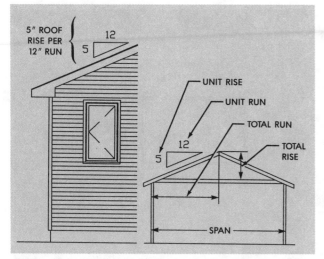

**Figure 5-4.** Roof slope is shown on elevation views with a slope symbol.

**Windows.** The window types most commonly used in residential construction are fixed-sash, double-hung, horizontal sliding-sash, casement, awning, and hopper. Elevation views show how the window sash is divided and where the hinges are located on a window that swings in or out. Dashed lines are used to show window swing. The apex of a triangle drawn on the window points to the side that is hinged. The triangle is drawn with dashed lines.

The size of the light may be shown on elevation views. A *light* is a pane of glass. For example, the sash of a window designated $^{28}/_{24}$ indicates that the glass size is 28″ wide and 24″ high. Note that the width is given first when designating window size.

Lines are omitted when window symbols are drawn. Wooden windows have wider sash parts than metal windows. Windows in framed walls have wood trim. Windows in masonry walls have a narrow brick mold. See Figure 5-6.

**Doors.** Flush and panel doors are commonly used in residential construction. *Flush doors* have flat surfaces of adjacent plies running at right angles to minimize movement and warpage. *Panel doors* have solid strips joined together to hold panels.

Flush doors are solid-core or hollow-core. *Solid-core doors* are made of solid blocks of wood or particleboard glued together and covered with veneer plies. These doors are commonly used for exterior openings. *Hollow-core doors* are made of wooden strips glued together on edge like an eggcrate and covered with veneer plies. Solid-core doors at least

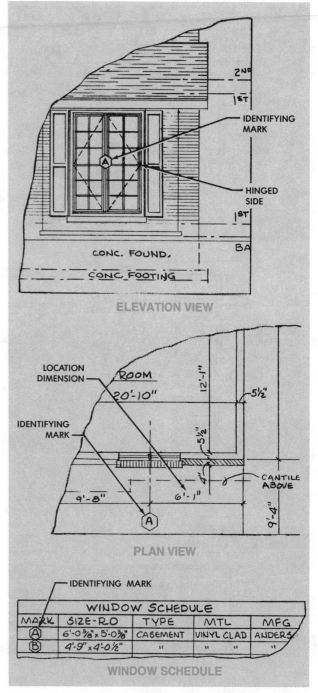

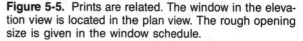

**Figure 5-5.** Prints are related. The window in the elevation view is located in the plan view. The rough opening size is given in the window schedule.

1¾″ thick are required for exterior openings. Three hinges are used because of the weight of the solid-core doors. Doors for interior openings must be at least 1⅜″ thick. Two hinges are required for interior doors.

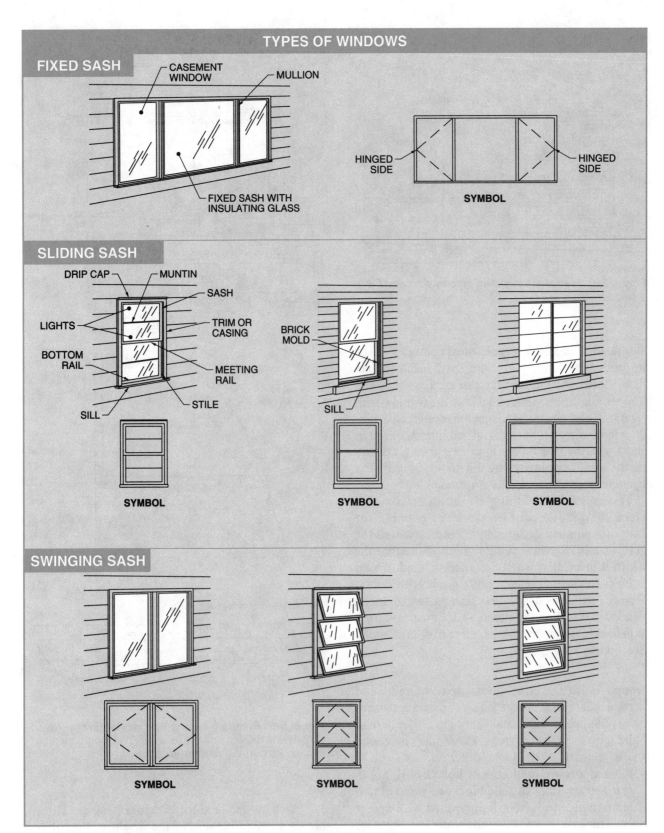

## TYPES OF WINDOWS

**FIXED SASH**

CASEMENT WINDOW

MULLION

HINGED SIDE

HINGED SIDE

**SYMBOL**

FIXED SASH WITH INSULATING GLASS

**SLIDING SASH**

DRIP CAP

MUNTIN

SASH

TRIM OR CASING

LIGHTS

BOTTOM RAIL

MEETING RAIL

SILL

STILE

BRICK MOLD

SILL

**SYMBOL**

**SYMBOL**

**SYMBOL**

**SWINGING SASH**

**SYMBOL**

**SYMBOL**

**SYMBOL**

**Figure 5-6.** Symbols on elevation views show the type of window used.

Metal doors may be flush or panel. They are insulated to help maintain inside temperatures. Lights, in a wide variety of patterns, make these very attractive for exterior openings. Little maintenance is required on metal doors.

*Door hand* is the direction a door swings. Hinges and hand are not shown on doors in elevation views. The floor plan is checked to determine the hinged side and hand. See Figure 5-7.

## Exterior Finish

Elevation views show the exterior finish of the house. Materials, trim, and miscellaneous details are shown with symbols and described by notations and dimensions. Wood, masonry, and siding of varying shapes and materials are shown. Shingles and other roof coverings are drawn as symbols on elevation views.

Elevation views also show decorative features designed to enhance the appearance of the house. These include columns, posts, balustrades, and decorative trim around doors, windows, and eaves. Weatherproof electrical receptacles, exterior lighting, frostproof hose bibbs, gutters and downspouts, and flashing are shown on elevation views. Refer to the Appendix.

## READING ELEVATION VIEWS

Plans for the Wayne Residence contain two sheets with exterior elevation views. Refer to Wayne Residence, Sheets 3 and 4. Sheet 3 shows the South and East Elevations. Sheet 4 shows the North and West elevations.

The title blocks show the name of the plans, scale, drafter, date, lot number and address, design company, and sheet numbers. The name of each elevation is given, with its scale, below each elevation view. The scale for all exterior elevations in this set of plans is $\frac{1}{4}'' = 1'\text{-}0''$.

## South Elevation

The South Elevation shows the front of the house. Refer to Sheets 1, 2, and 6 showing the symbol designating North which orients the house and names the elevations.

The dashed lines on the South Elevation show the concrete foundation footing and foundation wall. A $1'\text{-}0''$ diameter concrete pier provides additional

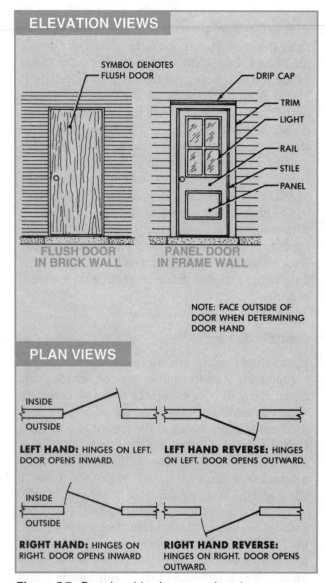

**Figure 5-7.** Door hand is shown on plan views.

support for the stoop. The foundation wall is stepped-down below the bedroom area of the house to provide a $6''$ ledge for the brick veneer. Vertical dimensions give measurements between finished floors and ceilings. The grade level is shown sloping to the East end of the house.

The South Elevation is finished with brick veneer. Casement windows, marked A and B, are shown. The apex of the dashed lines indicates the hinged side. Refer to the Window Schedule, Sheet 5 for additional information regarding windows. Dimensions from the house corners to the centers of the rough openings for these windows are shown on the Floor Plan, Sheet 1. Shutters flank the windows.

The entry door is a 3'-0" × 6'-8" × 1¾" exterior, metal-insulated, panel door. Refer to the Door Schedule, Sheet 5. One sidelight flanks the door. The center of the rough opening for the door and sidelight is 31'-3" from the framing for the Southeast corner of the house. It is 31'-7" from the finished Southeast corner. Refer to Sheets 1 and 2. A column on the stoop supports the roof overhang at the stoop.

The four-section overhead garage door is shown. For additional information regarding the garage door and its rough opening, refer to Sheets 1 and 2.

The hipped roof has a 6 in 12 slope. It is covered with cedar shakes and has a 1'-6" overhang. The chimney is shown rising above the roof. Notes specify brick bonds to be used. Refer to the Roof Plan, Sheet 5 for additional information regarding the roof and chimney.

## East Elevation

Foundation footings and walls are stepped-down at the retaining wall to compensate for the lot grade drop. The wall is brick veneer with two casement windows, one in Bedroom 1 and the other in the family room. The East wall at the rear of the house is finished with 12" lap siding. Openings in this wall show a full glass wood door for the living room and a casement window for the study. Refer to Sheets 1 and 2 and the Window and Door Schedules, Sheet 5.

Vertical dimensions give measurements between finished floors and ceilings. The deck wraps around the North wall of the house. Horizontal dimensions for the deck are shown on Sheet 2. Roof and chimney information is similar to that shown for the South Elevation.

## North Elevation

The North Elevation shows the rear of the house. Concrete piers are 1'-0" in diameter. They provide a base for the 4 × 4 CCA posts supporting the deck.

CCA is an acronym for chromated copper arsenate, a material used to treat lumber. The concrete foundation footing and foundation wall are stepped-down at the retaining wall to compensate for the lot grade drop.

The finish material on the rear wall is 12" lap siding. A vinyl-clad awning window is shown in the areaway of the future workshop. Refer to Sheet 1 and the Window Schedule, Sheet 5. Note that a break line is used so that the window may be shown with solid lines. A sliding glass door is shown for the family room. Refer to Sheet 1 and the Door Schedule, Sheet 5.

A wooden atrium door leads to the deck from Bedroom 1, and a wooden screen door is shown on the screened porch. Vinyl-clad casement windows are shown for the master bath and breakfast area. Fixed aluminum-framed patio replacement windows are shown for the living room. Refer to Sheet 1 and the Window Schedule, Sheet 5.

Information for the roof is similar to the other elevation views with the addition of seven louvers. These provide ventilation for the attic. They are placed on the rear of the house so they will not detract from the overall appearance when viewed from the street.

## West Elevation

The West Elevation shows the screened porch and garage end of the house. The concrete foundation footing and foundation wall are shown. The footing for the screened porch and garage is at the same level. The lower foundation footing relates to the future workshop. An end view of the portion of the deck that extends from the North side of the house is shown in the West Elevation. The wall is brick veneer with one vinyl-clad casement window for the laundry and one six panel, solid-core wooden door for the garage. Refer to Sheets 1 and 2 and the Window and Door Schedules, Sheet 5. Information for the roof is similar to the other elevation views.

# Sketching

Name _____ Date _____

## Sketching 5-1

**Sketch elevation symbols in the spaces provided. Refer to the Appendix.**

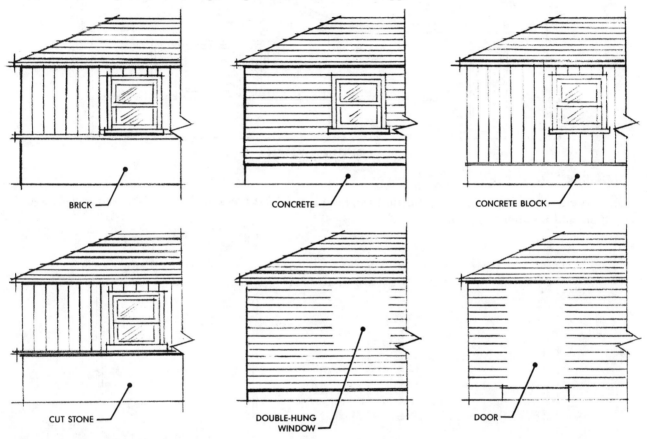

BRICK

CONCRETE

CONCRETE BLOCK

CUT STONE

DOUBLE-HUNG WINDOW

DOOR

## Sketching 5-2

**Complete the sketch to show windows hinged on the sides checked. Refer to the Appendix.**

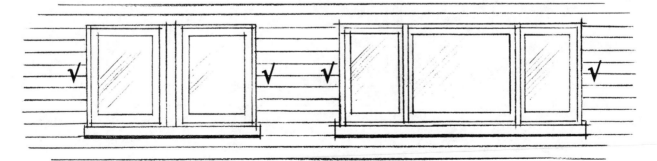

## Sketching 5-3

**Add symbols and notations as required to complete the sketch of the South Elevation.**

1. Roof slope is 5 in 12.
2. Cedar shake roof.
3. Full length shutters on both sides of windows and doors.
4. Brick veneer to 4'-0" above finished grade and lap siding above.
5. Sketch a casement window with mullion between the door and window shown.

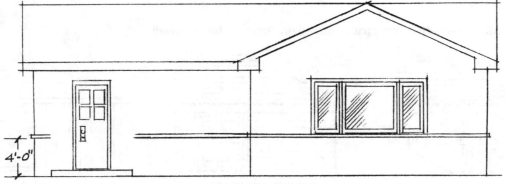

**SOUTH ELEVATION**

## Sketching 5-4.

**On separate sheets of paper, sketch North, East, South, and West Elevations of the Warner Residence. Add symbols and notations as required.**

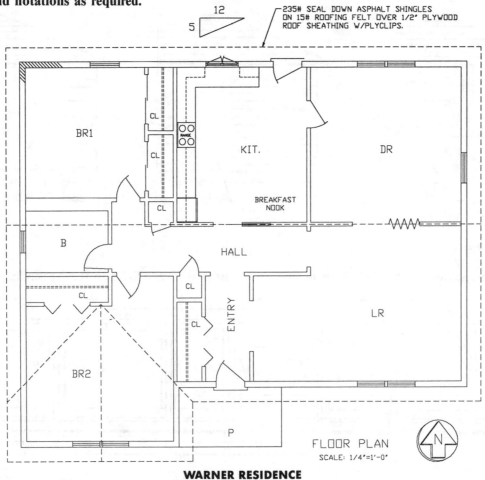

**WARNER RESIDENCE**

# Review Questions

Name _____ Date _____

## True-False

T    F    **1.** Elevations are pictorial drawings showing exterior views of a building.

T    F    **2.** Elevations show where exterior openings are placed.

T    F    **3.** Major points of a compass may be used to designate elevation views.

T    F    **4.** The various prints of a set of plans are related to each other.

T    F    **5.** Most elevations of residential construction are drawn to the scale of ⅛″ = 1′-0″.

T    F    **6.** Abbreviations in notations may be used with symbols to show building materials on elevations.

T    F    **7.** Abbreviations that form a word are followed by a period to distinguish them from the actual word.

T    F    **8.** A Cape Cod house has a traditional design.

T    F    **9.** A flat roof is required to slope at least ⅛″ per foot to provide for water runoff.

T    F    **10.** The slope of a roof is the relationship of the roof rise to the run.

T    F    **11.** Elevation views show horizontal dimensions for door and window openings.

T    F    **12.** The apex of a triangle drawn on a window in an elevation view points to the hinged side.

T    F    **13.** A window designated $^{32}\!/_{48}$ indicates that the glass is 48″ wide.

T    F    **14.** Flush doors may be either solid-core or hollow-core.

T    F    **15.** Three hinges are required for exterior doors in residential construction.

## Matching

_____    **1.** Gable roof                    A. Double slope in two directions

_____    **2.** Hip roof                      B. Double slope in four directions
                                                    C. Minimum slope of ⅛″ per foot
_____    **3.** Flat roof                      D. Single slope in two directions
                                                    E. Single slope in four directions
_____    **4.** Gambrel roof

_____    **5.** Mansard roof

## Multiple Choice

_____    **1.** The North Elevation is the _____.
              A. Elevation facing North
              B. direction a person faces to see the North side of the house
              C. either A or B
              D. neither A nor B

_____ 2. The scale of ¼″ = 1′-0″ indicates that the drawing is _____ th the size of the actual building.
- A. ¼
- B. ¹⁄₁₂
- C. ¹⁄₄₈
- D. ¹⁄₉₆

_____ 3. _____ roofs are popular on barns and country-style houses.
- A. Flat
- B. Hip
- C. Gambrel
- D. Mansard

_____ 4. In relation to roofs, run is the _____.
- A. vertical increase in roof height
- B. horizontal distance in which the roof rises
- C. overall distance between building corners
- D. overall distance, less space for brick veneer, between corners

_____ 5. The hinged side of a door can be determined from the _____.
- A. plot plan
- B. floor plan
- C. elevation views
- D. none of the above

_____ 6. Doors for exterior openings are required to be _____ thick.
- A. hollow-core and at least 1⅜″
- B. hollow-core and at least 1¾″
- C. solid-core and at least 1⅜″
- D. solid-core and at least 1¾″

_____ 7. The size of a window may be given _____.
- A. on the elevation views
- B. in a window schedule
- C. both A and B
- D. neither A nor B

_____ 8. Information commonly found on elevation views includes _____.
- A. dimensions for locating rough openings for doors and windows
- B. overall size and shape of the lot
- C. dimensions from established grade to finished first floor level
- D. size and location of joist spacing

_____ 9. Visible lines in elevation views are drawn as _____ lines.
- A. solid
- B. dashed
- C. intermittent long and short
- D. none of the above

_____ 10. When designing houses, drawings showing the _____ are generally drawn first.
- A. plot plan
- B. floor plan
- C. elevation views
- D. details

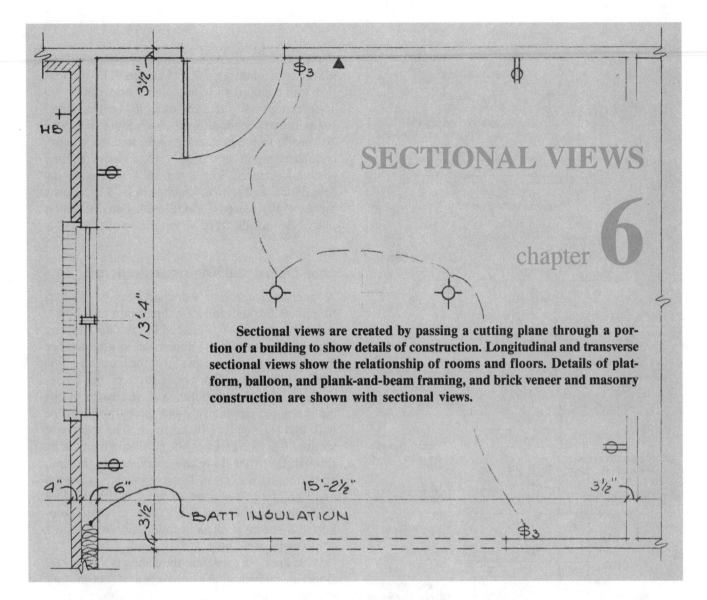

# SECTIONAL VIEWS

chapter **6**

Sectional views are created by passing a cutting plane through a portion of a building to show details of construction. Longitudinal and transverse sectional views show the relationship of rooms and floors. Details of platform, balloon, and plank-and-beam framing, and brick veneer and masonry construction are shown with sectional views.

## SECTIONAL VIEWS

Floor plans and elevation views give overall and location dimensions, show the material to be used in walls, partitions, and floors, and show the location of rough openings for doors and windows. Sectional views provide additional, detailed information. A *sectional view* is a scaled view created by passing a cutting plane vertically through a portion of a building. Sectional views drawn to a larger scale than the plan on which the cutting plane is drawn are known as detail views. Detail views show more detail because of their larger size.

Longitudinal and transverse sections are created by passing vertical cutting planes completely through the house to show the relationship of rooms and floors. The cutting planes for longitudinal and transverse sections are shown on the plan view.

## Typical Detail

A typical section is an enlarged sectional view that shows details of material and construction that are common to that part of the construction throughout the building. For example, a typical foundation footing and wall section shows, in detail, materials required and all necessary dimensions for constructing forms for the foundation footing and wall wherever they appear on the set of plans.

A typical wall detail is a sectional view taken by passing a cutting plane through the foundation footing and wall shown in the plan view. See Figure 6-1. The arrows indicate the direction of sight in which the section is taken. The detail of the section is identified as Detail 1 of Sheet 5. Additional information given on the enlarged Typical Wall Detail shows the size of the foundation footing and wall and the

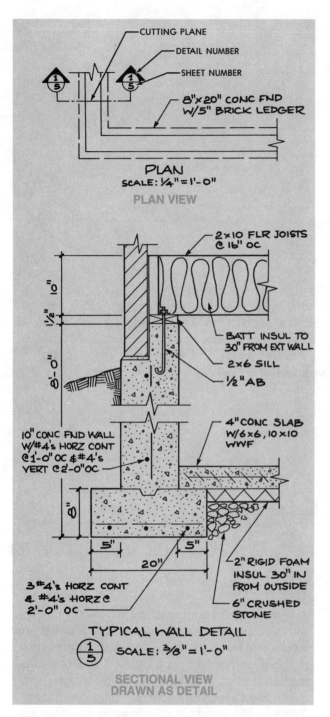

**Figure 6-1.** Sectional views are created by passing a cutting plane through a building feature to show the interior details of the feature. Details are sectional views drawn to larger scales.

size and spacing of rebars. The foundation footing is 8″ × 20″. The foundation wall is 10″ thick and 8′-0″ high including 8″ for the foundation footing. Three #4 rebars are placed horizontally throughout

the length of the foundation footing, with #4 rebars placed horizontally on 2′-0″ centers. The foundation wall contains #4 rebars placed horizontally on 1′-0″ centers and vertically on 2′-0″ centers. Rigid foam insulation beneath the concrete slab is placed 30″ inside from all exterior walls, and the 4″ concrete slab contains 6 × 6, 10 × 10 welded wire fabric for reinforcement. The 2 × 6 sill is secured to the foundation wall with ½″ anchor bolts. Floor joists are 2 × 10s placed 16″ OC with batt insulation placed 30″ inside from all exterior walls.

## Longitudinal and Transverse Sections

A *longitudinal section* is created by passing a cutting plane through the long dimension of a house. A *transverse section* is created by passing a cutting plane through the short dimension of a house. See Figure 6-2. The floor plan shows the location of the cutting plane lines that create the sections.

The longitudinal section, A-A, is created by the cutting plane passing through the bath, down the hall, and through the living room. The transverse section, B-B, is created by the cutting plane passing through the front door and entry, across the hall, and through the kitchen. The subfloor is ½″ plywood and 1 × 2 cross bridging reinforces the 2 × 10 floor joists placed 16″ OC. A W8 × 15 steel beam is supported by beam pockets in foundation walls and a 4″ diameter steel post.

Walls are ½″ drywall and the ceiling is ⅜″ drywall. Doors are referenced to a door schedule (not shown). Kitchen cabinets and appliances are shown on the North wall of the house. The countertop and 4″ backsplash are Micarta®. A 12″ soffit is above the wall cabinets. Batt insulation is placed between the 2 × 8 ceiling joists which are 16″ OC. Rafters are also 16″ OC. The exterior of the house is face brick, and the roofing is 235 lb asphalt shingles on 15 lb roofing felt over ½″ plywood sheathing.

## RESIDENTIAL CONSTRUCTION

Houses may be either frame construction, brick veneer construction, or solid masonry construction. The three general types of frame construction are platform (western), balloon, and plank-and-beam. A brick veneer building is also considered as frame construction because the internal supporting structure is wood framing, either platform or balloon.

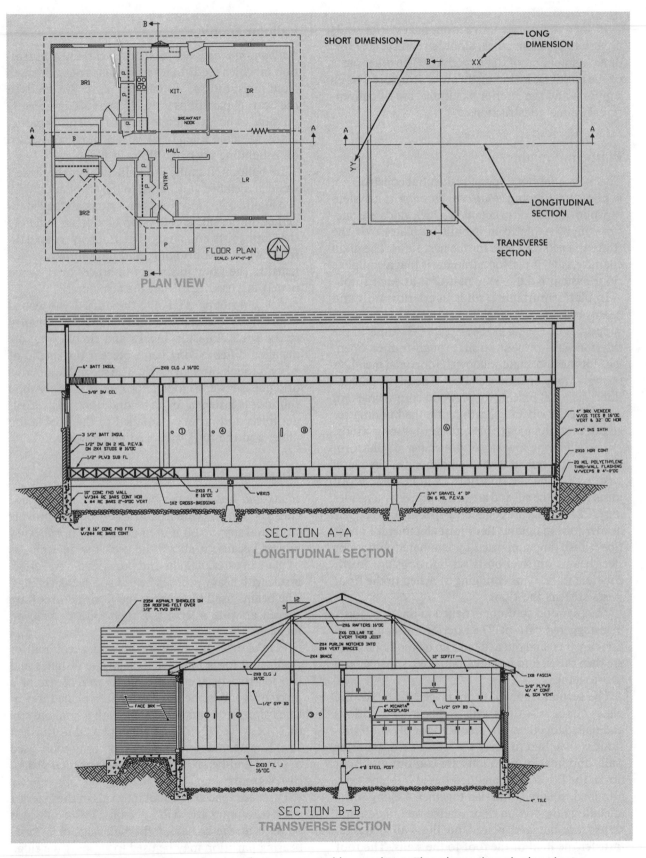

**Figure 6-2.** Longitudinal and transverse sections are created by passing cutting planes through plan views.

In solid masonry construction, the exterior walls are brick, concrete block, stone, structural clay tile, or a combination of these products. The other parts of the building, including the floor structure and its supports and the interior partitions, are of conventional frame construction.

## Platform Framing

The most common type of residential construction is platform framing. *Platform framing* is a system of wood-frame construction in which studs are one-story high. A platform is built on plates over the studs and acts as a base for the next floor. The main characteristic of platform framing is that a complete floor system is built as a platform at each floor.

In platform framing, studs in exterior walls and beams or girders through the center of the building support floor members. See Figure 6-3. Studs in the exterior walls provide strong, rough-framed openings for the doors and windows. Horizontal members support floor or roof members. The plywood subfloor, diagonal bracing (or plywood corner bracing), and the exterior wall sheathing tie the building together and provide stiffness to resist strong winds.

The sill is the lowest wood member of platform framing. It is bolted to the foundation wall in a level position so that the floor joists will be level. Floor joists rest on the sill and are tied together by a header joist. Floor joists are commonly spaced 16″ OC. The header joist maintains floor joist alignment. Longer floor joists are supported by laminated girders or steel beams on steel posts set into footings in the concrete floor. Cross bridging attached to the floor joists stiffens the floor.

The plywood subfloor is nailed to the floor joists in a staggered pattern. The exterior walls and partitions are then laid out and built up on the floor. They are then raised into position, plumbed, squared, and tied together.

The second floor joists are placed on top of the plates on the first floor walls and bearing partitions and nailed together. The second floor plywood subfloor is then laid to cover the whole area. Exterior walls and interior partitions are laid out and built up on the floor. They are then raised into position, plumbed, squared, and tied together. Ceiling joists are laid on the second floor double wall plates, and rafters are cut and placed on the wall plates to transfer the load of the roof to the walls. The roof is then sheathed and covered.

## Balloon Framing

*Balloon framing* is a system of wood-frame construction in which studs in exterior walls run full length from the sill to the double top plate. The studs at the bearing partitions are as long as is conveniently possible to help eliminate shrinkage. This is particularly valuable when brick veneer or stucco is used on a building over one floor in height.

In balloon framing, the sill is also the lowest wooden member. It is bolted to the foundation wall. See Figure 6-4. Floor joists and studs are nailed together on the sill, and firestop blocks are nailed to the floor joists and studs. The floor joists are stiffened with cross bridging, and the subfloor is fastened to the floor joists. Let-in braces or plywood panels are used to stiffen the structure.

A 1 × 4 ribbon is let in to the studs at the second floor. Floor joists rest on the ribbon and are nailed to the studs. Firestop blocks and fireblocking are installed. Fireblocking helps prevent the spread of a fire through the wall cavity. The second floor cross bridging is fastened to the floor joists. The plywood subfloor is fastened to the second floor joists. Ceiling joists and rafters are nailed to the double top plate, and the roof is decked and covered.

## Plank-and-Beam Framing

*Plank-and-beam framing* is a system of wood-frame construction in which planks and beams provide structural support. It is used extensively in one-story houses because it allows the use of wide expanses of glass in exterior walls and provides a novel effect inside with heavy exposed beams and plank ceilings. The beams used to support the floor and roof are spaced at wide intervals. Either solid or built-up members may be used.

Planks of 2″ nominal thickness are required for floors and roofs in order to span the distance and provide structural support. The *nominal size* of a piece of wood is the size before it is planed to finished size. For example, a plank with a nominal size of 2″ × 8″ is 1½″ × 7½″ when planed to finished size. Tongue-and-groove planks are commonly used in plank-and-beam framing because of their attractive V-groove.

Posts are spaced at equal intervals in the exterior walls to support the ceiling beams. Posts may also be used at the center of the building, or a load-bearing partition may be used to support the ridge beam. When posts are used in the exterior wall and

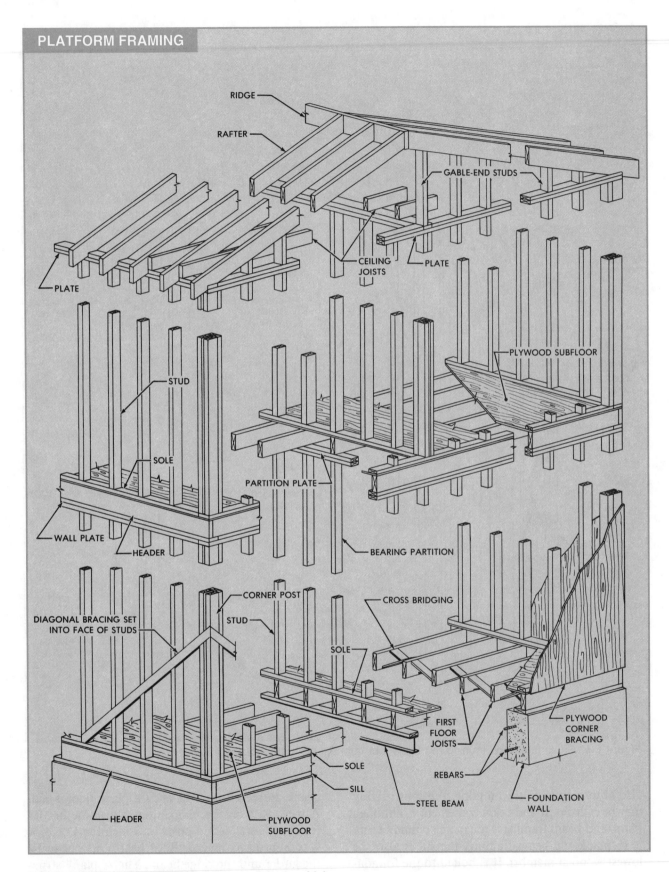

**Figure 6-3.** Studs in platform framing are one-story high.

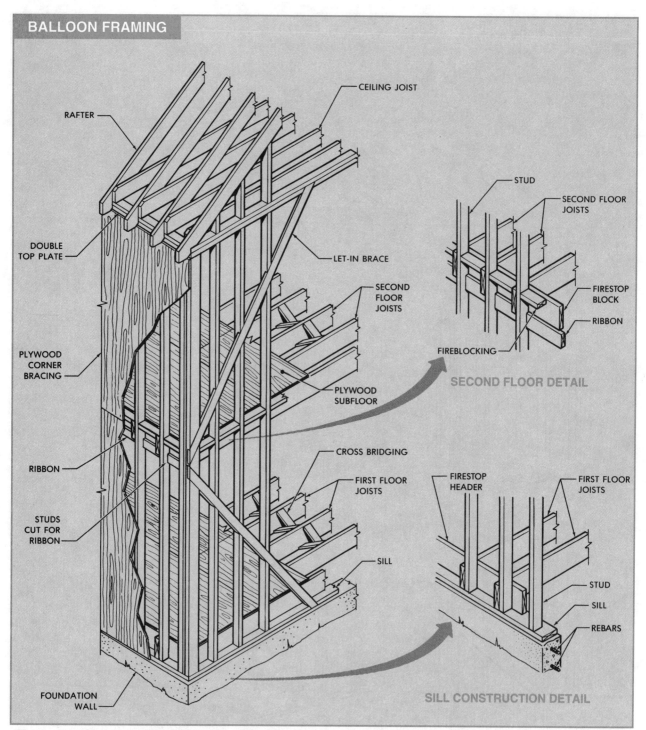

**BALLOON FRAMING**

RAFTER

CEILING JOIST

DOUBLE TOP PLATE

LET-IN BRACE

STUD

SECOND FLOOR JOISTS

SECOND FLOOR JOISTS

FIRESTOP BLOCK

RIBBON

PLYWOOD CORNER BRACING

FIREBLOCKING

PLYWOOD SUBFLOOR

**SECOND FLOOR DETAIL**

RIBBON

CROSS BRIDGING

STUDS CUT FOR RIBBON

FIRST FLOOR JOISTS

FIRESTOP HEADER

FIRST FLOOR JOISTS

SILL

STUD

SILL

REBARS

FOUNDATION WALL

**SILL CONSTRUCTION DETAIL**

**Figure 6-4.** Studs in balloon framing run full length from the sill to the double top plate.

also at the center of the building, the construction may be called post-and-beam construction, although plank-and-beam framing is the more common term.

In plank-and-beam framing, the sill is also the lowest wooden member. It is bolted to the foundation wall. See Figure 6-5. Header joists and floor beams are spiked together, and the plank floor is laid. A sole plate on the plank floor above the header joists supports 4 × 4 posts on which the 4 × 6 or 4 × 8 girder is placed. Roof beams are cut to fit the girder and the ridge beam. The 2″ plank structural roof is then applied.

## PLANK-AND-BEAM FRAMING

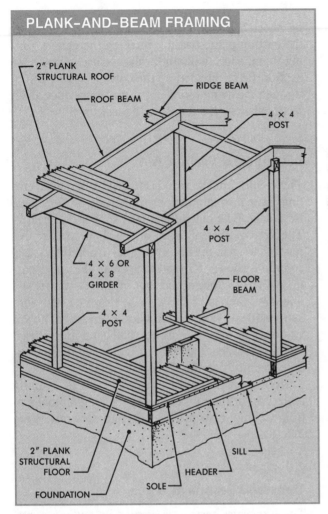

**Figure 6-5.** Planks and beams provide structural support in plank-and-beam construction.

### Brick Veneer Construction

*Brick veneer construction* is frame construction with a brick wall or other masonry units. Brick veneer walls may be laid with platform or balloon framing. An air space is maintained between the brick and the sheathing to permit a slight amount of shifting between the two walls. Metal ties are used to attach the brick veneer wall to the frame construction. Weep holes at the base of the brick veneer wall allow moisture to escape.

The foundation wall for brick veneer construction is offset to provide a base for the brick veneer. See Figure 6-6. Metal base flashing is extended up at least 6″ behind the sheathing paper to prevent moisture penetration into the wooden structure. Metal ties are nailed to the studs and embedded in the mortar. The balance of the construction is platform framing.

### Masonry Construction

*Masonry construction* has masonry units such as brick, concrete block, stone, or structural clay tile that form walls to carry the load of floor and roof joists. The joists are fire cut to rest in the solid brick wall. A *fire cut* is an angled cut in the end of a joist that allows a burnt joist to fall out without disturbing the solid brick wall. Joists are secured to the wall with joist anchors designed to pull out of the wall in case of fire.

A foundation footing and wall sufficient to carry the load are the base for solid brick walls in masonry construction. See Figure 6-7. The first floor joists rest on a wythe of brick. A *wythe* is a single, continuous masonry wall one unit thick. The balance of the solid masonry wall is two bricks thick. Upper-story, ceiling, and roof joists are fire cut. Load-bearing partitions through the center of the building are platform framed.

## READING SECTIONAL VIEWS

Interior spaces of the Wayne Residence are shown on the sectional views. See Wayne Residence, Sheet 7. Cutting planes for the sectional views are shown

## BRICK VENEER CONSTRUCTION

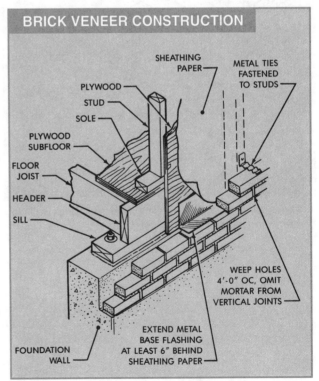

**Figure 6-6.** Brick veneer construction is frame construction with a brick wall or other masonry units.

## MASONRY CONSTRUCTION

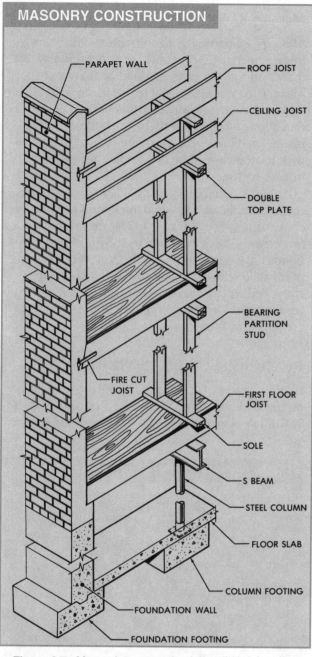

PARAPET WALL

ROOF JOIST

CEILING JOIST

DOUBLE TOP PLATE

BEARING PARTITION STUD

FIRE CUT JOIST

FIRST FLOOR JOIST

SOLE

S BEAM

STEEL COLUMN

FLOOR SLAB

COLUMN FOOTING

FOUNDATION WALL

FOUNDATION FOOTING

**Figure 6-7.** Masonry construction has solid brick walls to carry the load of floor and roof joists.

on Sheet 2. Longitudinal Section A-A is created by the cutting plane passing through the garage, dining room, entry, hall, and walk-in closet. Transverse Section B-B is created by the cutting plane passing through the living room, entry, and entry door. Both sectional views are drawn to the scale of $\frac{1}{4}'' = 1'-0''$.

### Longitudinal Section A-A

The longitudinal section is created by passing a cutting plane through the long dimension of the Wayne Residence. The cutting plane is shown on Sheet 2. The W8 × 10 steel beam on the 3" diameter steel post is shown in the family room. An additional steel column is shown in the storage area. The upper portion of the #6 door at the foot of the stairway is visible. Two shelves in the walk-in closet of Bedroom 3 are made of 1x material. Clothes rods are shown. Doors to the master bedroom, hall bath, linen closet, and the door leading from the garage to the kitchen are referenced to the Door Schedule.

Floor-to-ceiling clearances and sizes of framing materials for the roof complete the longitudinal section. Ceiling joists are 2 × 10s and roof rafters are 2 × 8s.

### Transverse Section B-B

The transverse section is created by passing a cutting plane through the short dimension of the Wayne Residence. The cutting plane is shown on Sheet 2. The Stair Detail, also shown on Sheet 7, is directly related to the stairway as seen in this sectional view. A wrought iron handrail runs from the hallway to the living room.

The 4" column on the front entry is supported by an 18" concrete pier. Doors and windows are referenced to the schedules, and floor-to-ceiling heights are given.

# Sketching

Name _____ Date _____

## Sketching 6-1

**Sketch a typical section of a platform-framed house from the concrete foundation footing through 24″ above the finished first floor in the space provided. Use the scale ¾″ = 1′-0″. Letter all parts of the sketch.**

1. 10″ × 20″ foundation footing w/2″ × 4″ keyway and 3 #4s horizontal continuously and #4s horizontal at 2′-0″ OC

2. 4″ clay drain tile

3. 10″ × 36″ foundation wall to 6″ above grade level with #4s horizontal continuously at 1′-0″ OC and #4s vertical at 2′-0″ OC

4. Termite shield

5. 2 × 6 sill with ½″ AB

6. 2 × 10 header

7. 2 × 10 floor joists with 1 × 2 cross bridging

8. ½″ plywood subfloor

9. ¾″ oak tongue-and-groove finished floor

10. 2 × 4 sole

11. 2 × 4 studs

12. ½″ exterior sheathing

13. 8″ lap siding

14. ½″ drywall

15. R-13 batt insulation in walls

## Sketching 6-2

**Sketch a typical section of a balloon-framed house from the concrete foundation footing through 24" above the finished first floor in the space provided. Use the scale ¾" = 1'-0". Letter all parts of the sketch.**

1. 10" × 20" foundation footing w/ V-shaped keyway and two #4s horizontal continuously and #4 horizontal at 2'-0" OC

2. 4" PVC drain tile

3. 10" × 30" foundation wall to 4" above grade level with #4s horizontal continuously at 1'-0" OC and #4s vertical at 2'-0" OC

4. Termite shield

5. 2 × 6 sill with ½" AB

6. 2 × 10 floor joists with solid bridging

7. Batt insulation to 30" in from exterior wall

8. 2 × 10 firestop blocks

9. ¾" plywood subfloor

10. VA tile finished floor

11. 2 × 4 studs

12. ½" exterior sheathing

13. 4" lap siding

14. ½" drywall

15. R-13 batt insulation in walls

# Review Questions

Name _____    Date _____

## Identification 6-1

| | |
|---|---|
| _____ | 1. Floor joist |
| _____ | 2. Sill |
| _____ | 3. Cross bridging |
| _____ | 4. Wall double top plate |
| _____ | 5. Fireblocking |
| _____ | 6. Ribbon |
| _____ | 7. Solid bridging |
| _____ | 8. Partition stud |
| _____ | 9. Ceiling joist |
| _____ | 10. Rafter |
| _____ | 11. Steel beam |
| _____ | 12. Foundation wall |
| _____ | 13. Partition double top plate |
| _____ | 14. Steel post |
| _____ | 15. Rebar |

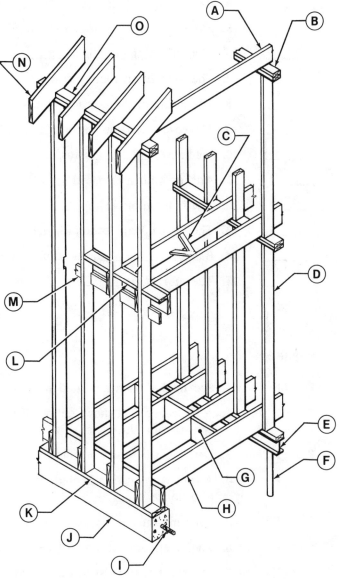

## Completion

**Refer to Garner Residence—Typical Wall Section on page 111.**

_____ 1. Ties for brick veneer are spaced _____″ OC vertically.

_____ 2. The concrete foundation wall measures _____ from the top of the foundation footing to the bottom of the sill.

_____ 3. Roof trusses are placed _____″ OC.

_____ 4. R-_____ batt insulation is placed in exterior walls.

_____ 5. Floor support is provided by 12″ D top-bearing floor _____.

_____ 6. Metal flashing extends _____ below the bottom wythe of brick.

_____ 7. The foundation footing contains _____ continuous #4 rebars.

_____ 8. The inside of the exterior wall is finished with _____.

_____ 9. The eave overhang measures _____ at 8/12 slope.

_____ 10. Medium _____ shingles are applied to the roof.

_____ 11. Anchor bolts are spaced _____″ OC in the foundation wall.

_____ 12. The foundation wall is _____″ wide.

_____ 13. The Typical Wall Section for the Garner Residence is drawn to the scale of _____.

_____ 14. Batt or _____ insulation may be used above the ceiling.

_____ 15. The subfloor is _____″ plywood.

## True-False

T    F    1. Sectional views are created with cutting planes.

T    F    2. A typical section shows details of material and construction common to that type of construction throughout the building.

T    F    3. Detail 2/6 is found on Sheet 2.

T    F    4. A transverse section is made by passing a cutting plane through the long dimension of a house.

T    F    5. Solid masonry is a type of frame construction.

T    F    6. The most common type of residential construction is platform framing.

T    F    7. The sill is the lowest wooden member in platform framing.

T    F    8. Studs run full length from the sill to the double top plate in balloon framing.

T    F    9. Solid or built-up beams may be used in plank-and-beam construction.

T    F    10. The nominal size of a piece of wood is its size after planing.

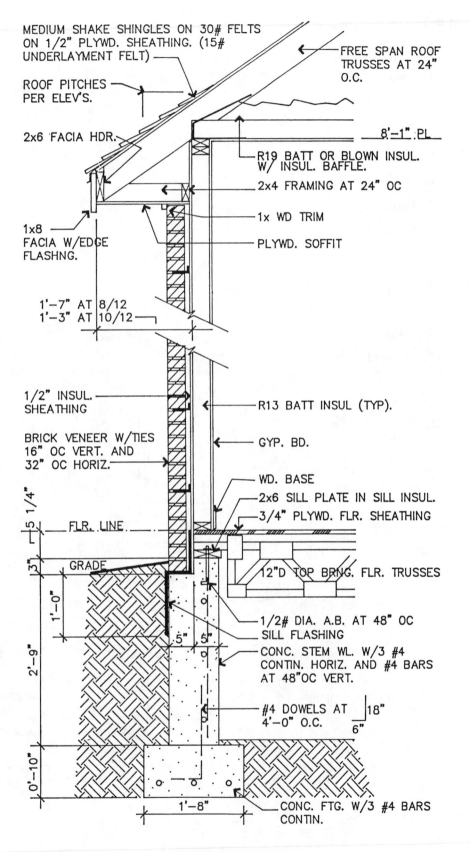

MEDIUM SHAKE SHINGLES ON 30# FELTS
ON 1/2" PLYWD. SHEATHING. (15#
UNDERLAYMENT FELT)

FREE SPAN ROOF
TRUSSES AT 24"
O.C.

ROOF PITCHES
PER ELEV'S.

2x6 FACIA HDR.

8'-1" PL

R19 BATT OR BLOWN INSUL.
W/ INSUL. BAFFLE.

2x4 FRAMING AT 24" OC

1x WD TRIM

1x8
FACIA W/EDGE
FLASHNG.

PLYWD. SOFFIT

1'-7" AT 8/12
1'-3" AT 10/12

1/2" INSUL.
SHEATHING

R13 BATT INSUL (TYP).

GYP. BD.

BRICK VENEER W/TIES
16" OC VERT. AND
32" OC HORIZ.

WD. BASE

2x6 SILL PLATE IN SILL INSUL.

3/4" PLYWD. FLR. SHEATHING

5 1/4"

FLR. LINE

12"D TOP BRNG. FLR. TRUSSES

3"

GRADE

1'-0"

1/2# DIA. A.B. AT 48" OC

SILL FLASHING

5" 5"

CONC. STEM WL. W/3 #4
CONTIN. HORIZ. AND #4 BARS
AT 48"OC VERT.

2'-9"

#4 DOWELS AT
4'-0" O.C.

18"

6"

0'-10"

1'-8"

CONC. FTG. W/3 #4 BARS
CONTIN.

**GARNER RESIDENCE**

## Identification 6-2

_____    **1.** Foundation wall

_____    **2.** Diagonal bracing

_____    **3.** Rafter

_____    **4.** First floor joist

_____    **5.** Sill

_____    **6.** Double top plate

_____    **7.** Partition plate

_____    **8.** Ceiling joist

_____    **9.** Plywood subfloor

_____    **10.** Rebar

_____    **11.** Header

_____    **12.** Ridge

_____    **13.** Plywood corner bracing

_____    **14.** Stud

_____    **15.** Cross bridging

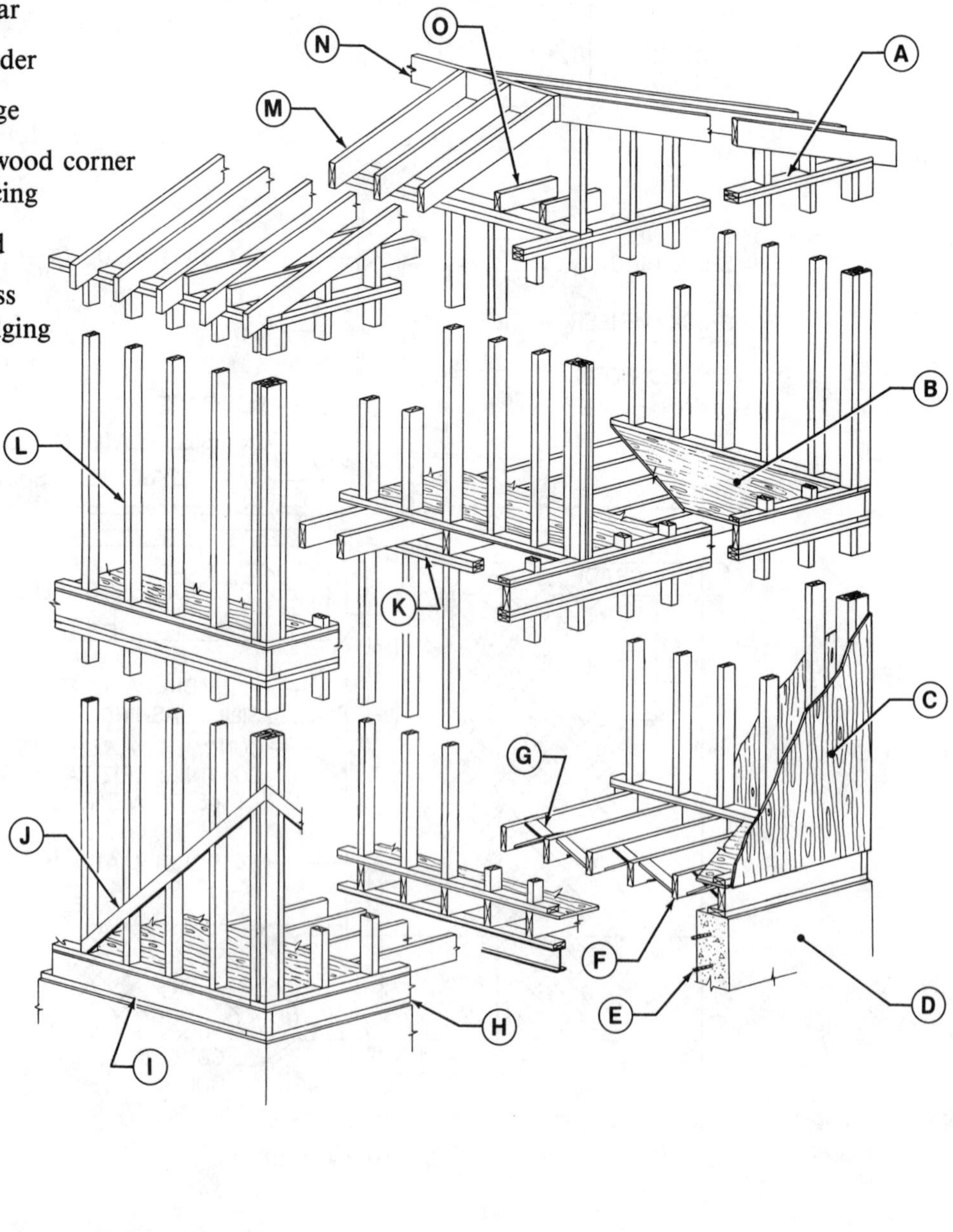

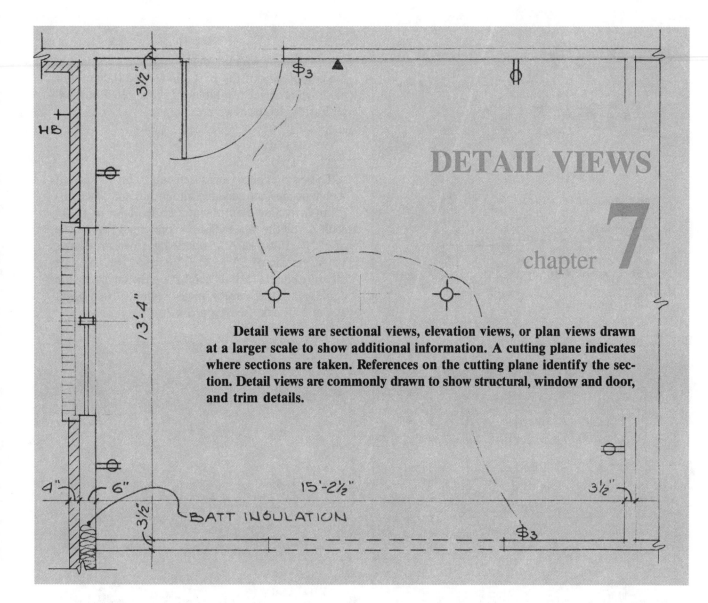

3½"

HB

13'-4"

4"   6"

3½

BATT INSULATION

15'-2½"

3½"

$_3

$_3

**Detail views are sectional views, elevation views, or plan views drawn at a larger scale to show additional information. A cutting plane indicates where sections are taken. References on the cutting plane identify the section. Detail views are commonly drawn to show structural, window and door, and trim details.**

## DETAILS

Architects try to show all the graphic information required by tradesworkers on plan views and elevation views. Specifications give additional written explanations and descriptions. Some features, however, must be drawn at a larger scale because sufficient information cannot be shown at the smaller scale of plan and elevation views.

*Detail views* are sectional, elevation, or plan views that are drawn to show more information. Details are *generally* drawn to a larger scale, although not always. Prints for residential construction commonly include details of interior wall elevations, structural parts and special features, windows and doors, and exterior and interior trim. See Figure 7-1.

## Scales

Details are usually drawn at a larger scale than the view from which they are taken. For example, details of sectional views showing structural information such as footings and walls, floor systems, and stairways are drawn at a larger scale, as are details of special features such as fireplaces, stair trim, and windows. Details of exterior trim showing cornices, dormers, and front entryways and interior trim such as cabinets, bookcases, mantels, paneling, and wall trim are also drawn at a larger scale.

An elevation detail of a plan view may be drawn at the same scale as the plan view. The use of the same scale allows dimensions to be transferred directly from the plan view to the elevation detail.

## DETAIL VIEWS

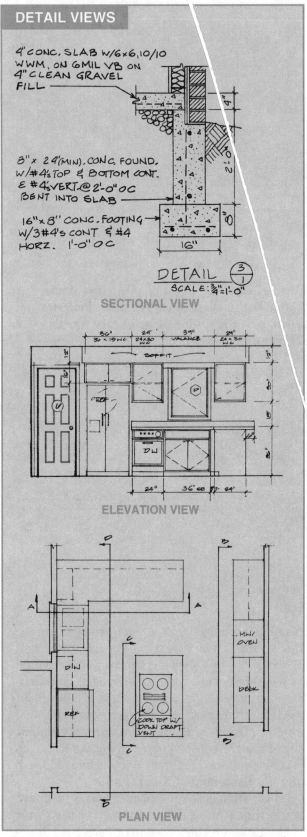

4" CONC. SLAB W/6x6,10/10
WWM, ON 6MIL VB ON
4" CLEAN GRAVEL
FILL

8" x 24"(MIN), CONC. FOUND.
W/#4's TOP & BOTTOM CONT.
& #4's VERT. @ 2'-0" OC
BENT INTO SLAB

16" x 8" CONC. FOOTING
W/3 #4's CONT & #4
HORZ. 1'-0" OC

DETAIL (3/1)
SCALE: ¾"=1'-0"

SECTIONAL VIEW

ELEVATION VIEW

PLAN VIEW

**Figure 7-1.** Detail views are sectional views, elevation views, or plan views drawn to show more information.

The scale at which a detail is drawn depends on the complexity of the part shown and the amount of drawing space available on the sheet. Preferred scales for detail views are:

¾" = 1'-0"
1½" = 1'-0"
3" = 1'-0"

These scales allow measurements to be worked out with tape measures divided into inches and sixteenths of an inch. For example, at a scale of ¾" = 1'-0", each ¹⁄₁₆" on the tape measure represents 1". At a scale of 1½" = 1'-0", each ⅛" on the tape measure equals 1". At a scale of 3" = 1'-0", each ¼" on the tape measure equals 1" and each ¹⁄₁₆" on the tape measure equals ¼". For accuracy, measurements often begin at the 1" mark. See Figure 7-2.

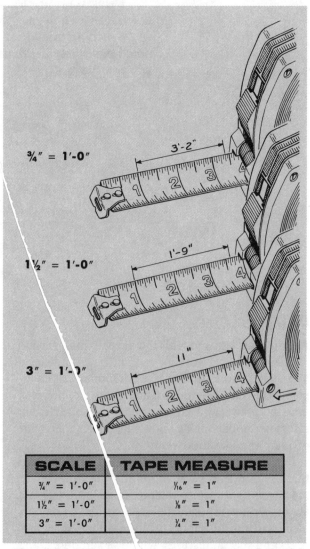

| SCALE | TAPE MEASURE |
|---|---|
| ¾" = 1'-0" | ¹⁄₁₆" = 1" |
| 1½" = 1'-0" | ⅛" = 1" |
| 3" = 1'-0" | ¼" = 1" |

**Figure 7-2.** A tape measure may be used to determine dimensions of detail views.

Full-size details are rarely drawn for residential construction unless the part is very small or very complex. For example, full-size details may be shown of small parts such as moldings or complex parts such as stair handrails.

## Dimensions

Dimensions are shown on detail views if they are not shown elsewhere on the plan and elevation views. Generally, dimensions are repeated only when necessary for clearness. If the same feature were to be dimensioned in different views with different dimensions applied, confusion would result. When dimensions are shown only once, it may be necessary to refer to several prints to determine the dimension.

The two basic types of dimensions are location dimensions and size dimensions. *Location dimensions* locate a particular feature in relation to another feature. For example, a dimension showing the distance from a building corner to the center of the rough opening for a window is a location dimension. *Size dimensions* give the size of a particular area or feature. For example, dimensions showing the overall size of the house and dimensions showing the height and width of foundation footings are size dimensions.

Architects place dimensions on the appropriate view for what is shown. It is often necessary to refer to several prints to find all dimensions related to a particular feature.

## References

Details are referenced to the view from which they are taken by cutting plane lines, direction of sight symbols, and notations. These references indicate that more information is available elsewhere on the same sheet or on another sheet of the prints. A detail is placed on the same sheet as its reference, if possible. Details may also be placed together on a separate sheet from their references.

**Cutting Planes.** Cutting planes are imaginary slices passed through a feature to show its cross section and give additional detail. A *cutting plane line* is drawn to show the exact place where the cutting plane passes through the feature. Arrows on the cutting plane line indicate the direction of sight in which the sectional view is taken. When several sectional views are taken through walls, foundations, or other parts of the building, they are designated SECTION A-A, SECTION B-B, and so forth.

**Notations.** A notation such as SEE DETAIL is often used to indicate that additional information concerning the referenced feature is shown elsewhere. Symbols with notations may be used to indicate the direction of sight for interior elevation details. The elevation detail number and the sheet number on which the detail is drawn are shown on the symbol. For example, a symbol with the notation 1/5 refers to DETAIL 1, SHEET 5. A symbol with the notation 2/5 refers to DETAIL 2, SHEET 5, and so forth. Sectional details from cutting planes may be named by the same method. For typical details, no specific reference is required. The detail is noted as a typical detail by a notation such as TYPICAL WALL SECTION. See Figure 7-3.

## Interior Wall Elevations

Details of interior wall elevations are often drawn to provide more information than is shown on the plan view. For example, the plan view of a kitchen shows the location of base and wall cabinets, space for the sink and built-in appliances such as dishwashers, compactors, and wall ovens, and space for movable appliances such as refrigerators and ranges. Additional information on the plan view shows the location of windows, doors, and structural parts of the dwelling along with electrical receptacles and outlets. Often, however, tradesworkers need additional information to build and finish soffits, cabinets and countertops, and to install lighting, hoods, fans, and other features.

The plan view of a kitchen and its elevation details are related. See Figure 7-4. The plan view shows that the kitchen and breakfast room measure 11'-8" × 19'-4". L-shaped base and wall cabinets are placed on the South and East walls, and an island cabinet containing a range and grill separates the kitchen and breakfast room. The location of appliances, electrical receptacles, telephone outlet, windows, broom closet, and the cased opening to other rooms of the house are also shown on the floor plan. The three elevations of the kitchen provide additional information such as the pairing of cabinet doors, Formica® countertop and full backsplash, paneled broom closet doors, and a brick surround on the island cabinet with a copper hood above. Elevation dimensions for the cabinets are also given.

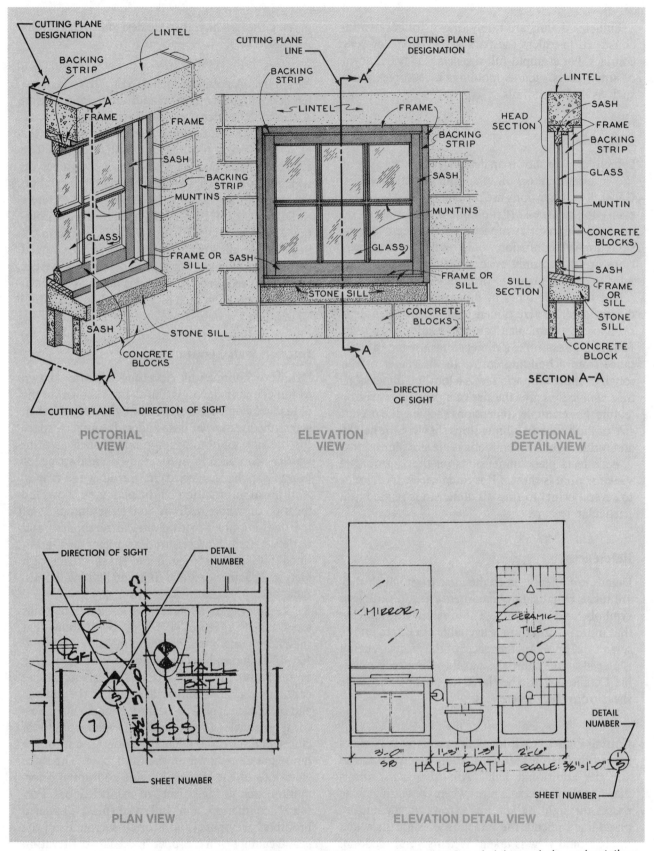

PICTORIAL VIEW

ELEVATION VIEW

SECTIONAL DETAIL VIEW

SECTION A-A

PLAN VIEW

ELEVATION DETAIL VIEW

HALL BATH    SCALE: 3/8"=1'-0"

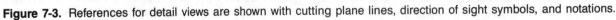

**Figure 7-3.** References for detail views are shown with cutting plane lines, direction of sight symbols, and notations.

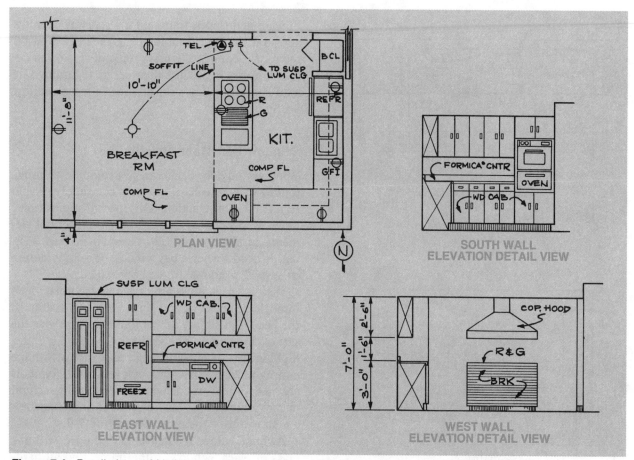

**Figure 7-4.** Detail views of kitchen elevations supplement information shown on the floor plan.

The plan view of a bath and its elevation details are related. See Figure 7-5. The plan view contains a bath with a separate 5'-0" × 6'-0" powder room. By relating the two views, a tradesworker can see that the double-bowl vanity measures 30" × 6'-0" and has three all-drawer units. GFI receptacles are located in walls on both ends of the vanity. A 4'-0" × 5'-11" mirror rests on the 6" vanity backsplash and extends to the 12" soffit. Two #3 doors are shown. The closet in the powder room has a pocket door which is identified as a #5 door. Additional information concerning the size of doors is given in a Door Schedule (not shown). The closet is illuminated by a pull-switch light fixture. Shelves in the powder room closet extend from floor to ceiling.

The tub and water closet are located in a 5'-0" × 6'-0" room with ceramic tile from floor to soffit. The tub has a sliding door. The soffit is the same size as the soffit in the powder room. A 36" × 42" wall cabinet is hung above the water closet and extends to the soffit.

Illumination is provided by ceiling lights controlled by wall switches. A wall switch also controls the ceiling fan above the water closet. The floor for the bath is ceramic tile.

Other walls throughout a house may have special treatments that also require elevations to show greater detail. For example, fireplaces, bookcases, and special window trim may be shown on elevation views.

## Structural Details

A main sectional view of the foundation footings and walls provides structural information needed to build a house. Frequently, more structural information than is shown in the one main sectional view is required. This occurs when the foundation footings and walls are formed differently to meet special conditions such as grade slopes and varying load requirements. Details are drawn to provide the additional information. Each of these is labeled with corresponding letters on the plan view and details. Arrows

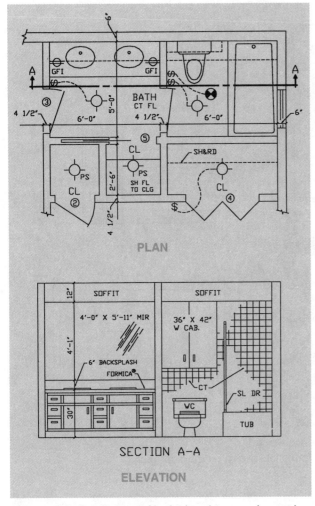

**Figure 7-5.** Detail views of bath elevations supplement information shown on the floor plan.

on the plan view show the direction of sight to make the cut to provide the detail. See Figure 7-6.

Fireplace details cannot be shown adequately on plan views and small scale elevation views. Details regarding the shape and size of the fireplace are required for construction. The arrangement of the damper, shape and size of the transition to the chimney, and size of the firebox must conform to the dimensions given on the plans for the fireplace to have efficient combustion and draft. The hearth is designed so that there is no danger of transmitting heat to the floor joists. See Figure 7-7.

Rough framing details for openings in walls and floors are generally worked out by carpenters. However, if structural design is involved and loads must be supported in certain ways, the architect provides details. This may include details showing framing

of walls and floors, framing for a stairway opening, and so forth. Roof framing plans are only shown if the roof is complex. Trusses for roofs are often detailed, however, because their design regarding safe load characteristics is the architect's responsibility. See Figure 7-8.

## Window and Door Details

Window and door details are not common on prints today as ready-to-hang units are commonly used in residential construction. For special applications, however, details are drawn to show structural elements, installation procedures, and trim. Such a situation occurs when a bay window or a combination of several windows are to be installed.

Details of windows are drawn as sectional views through the head, sill, and jamb. The sections through the head and sill are drawn looking directly at the side of the window. The section of the jamb is drawn looking directly down through the window with the cutting plane taken slightly above the sill. Symbols are used to show wooden members. Rough structural members are shown with a cross, and finish members are shown as woodgrain. See Figure 7-9.

Awning widows are delivered to the job with the glass in place, window in the frame, and all hardware installed. Dimensional allowances are given so that the carpenter can provide the proper size opening. See Figure 7-10. The section in the jamb is revolved 90° and placed in line with the sections through the head and sill. The sections through the head and the jamb show the difference between the top and side pieces of the sash and the top and side pieces of the window frame. (*Note:* Cover the section through the jamb to see the proper relationship between the sections through the head and sill.)

## Trim Details

Details of exterior and interior trim are often drawn in large scale views to clearly show the materials used. Exterior details are commonly drawn of siding, dormers, doorways, and unusual brick patterns. Interior details are commonly drawn of cabinets, stairways, millwork, and trim.

A detail of a cornice provides framing and finish information. See Figure 7-11. The wall is framed with 2 × 4 studs and filled with 3½″ batt insulation. Ceiling joists are 2 × 10s filled with 6″ batt insulation. Rafters are 2 × 8s covered with ½″ plywood and a

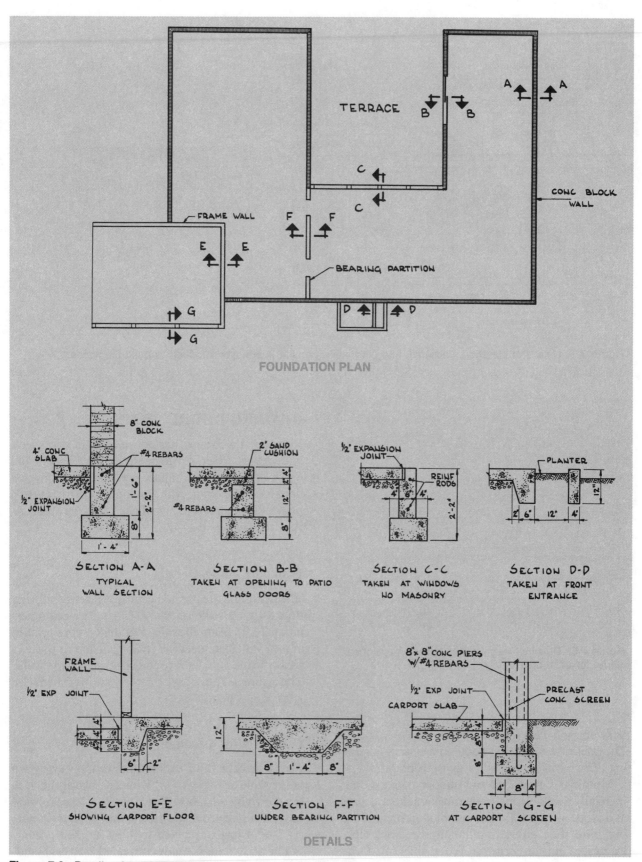

FOUNDATION PLAN

TERRACE

CONC BLOCK WALL

FRAME WALL

BEARING PARTITION

**SECTION A-A**
TYPICAL WALL SECTION

8" CONC BLOCK
4" CONC SLAB
#4 REBARS
½" EXPANSION JOINT
1'-6"
2'-2"
8"
1'-4"

**SECTION B-B**
TAKEN AT OPENING TO PATIO GLASS DOORS

2" SAND CUSHION
#4 REBARS
4"
12"
8"

**SECTION C-C**
TAKEN AT WINDOWS NO MASONRY

½" EXPANSION JOINT
REINF RODS
4"
6"
4"
2'-2"

**SECTION D-D**
TAKEN AT FRONT ENTRANCE

PLANTER
12"
2"
6"
12"
4"

**SECTION E-E**
SHOWING CARPORT FLOOR

FRAME WALL
½" EXP JOINT
6"
2"

**SECTION F-F**
UNDER BEARING PARTITION

12"
8"
1'-4"
8"

**SECTION G-G**
AT CARPORT SCREEN

8"x 8" CONC PIERS W/#4 REBARS
½" EXP JOINT
CARPORT SLAB
PRECAST CONC SCREEN
8"
4"
8"
4"

DETAILS

**Figure 7-6.** Details of sectional views provide structural information.

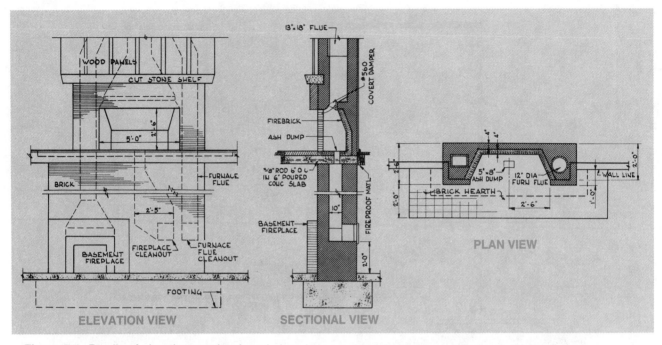

**Figure 7-7.** Details of elevation, sectional, and plan views provide information required to build fireplaces.

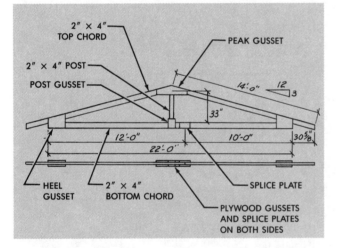

**Figure 7-8.** Detail views of trusses show dimensions required for safe loads.

built-up roof. They are trimmed with a 1 × 10 fascia. The exterior wall is ¾" plywood siding over ½" sheathing. The interior wall is ½" gypsum board.

Moldings for exterior and interior trim are commercially available in a wide range of shapes, sizes, and types of wood. Common moldings include casing, base, shoe, crown, mullion, apron, brick, cove, inside and outside corner, stop, and chair rail. Other moldings are also available. See Appendix.

## READING DETAIL VIEWS

Plans for the Wayne Residence contain two sheets with details. Refer to Wayne Residence, Sheets 5 and 6. The Kitchen Plan, Sheet 6 shows no additional details of the floor plan but does provide cutting planes showing directions of sight for Elevations A-A, B-B, C-C, and D-D.

### Panned Ceiling Detail

The detail for the panned ceiling is a sectional view drawn to the scale of ¾" = 1'-0". The ceiling extends 2'-0" from all walls and then slopes upward at a 45° angle to meet the center portion of the ceiling which is 1'-0" above the ceiling near the walls. The ceiling is framed with 2 × 4s and finished with ½" gypsum board.

### Deck Handrail Detail

The Deck Handrail Detail is an elevation view drawn at the scale of ½" = 1'-0". Dimensions for the deck are shown on the Floor Plan, Sheet 2. The detail view provides dimensions for the handrail. The deck rests on 2 × 8 joists attached to 4 x 4 treated posts.

The deck floor is made of treated 2 × 6s. Handrail uprights are treated 2 × 4s spaced 6" OC. These

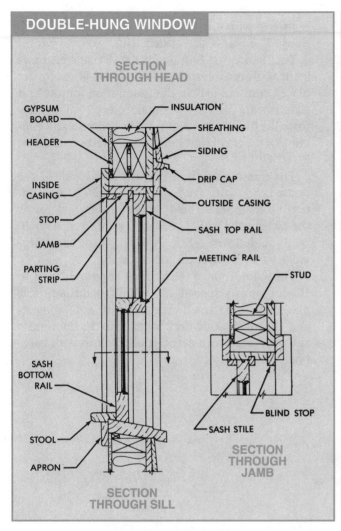

**Figure 7-9.** Details of sectional views through the head, sill, and jamb show construction information for windows.

**AWNING WINDOW**

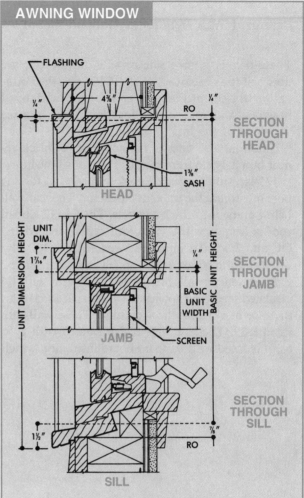

**Figure 7-10.** Details of sectional views through the head, sill, and jamb of a window are related.

are mitered at the top and bottom. Treated 2 × 4s and 2 × 6s form the handrail which is 3'-2" above the deck floor.

### Fireplace Details

An elevation view and a sectional view of the brick fireplace are drawn as detail views at the scale of ½" = 1'-0". The elevation detail shows marble surrounding the firebox with a 2 × 10 mantel 6'-0" long. The fireplace is 6'-0" wide.

The sectional view of the fireplace shows a ceramic tile hearth extending 1'-6" from the firebox opening. Brick symbols show firebrick around the firebox and face brick elsewhere.

**CORNICE**

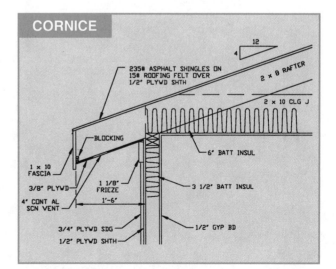

**Figure 7-11.** Details of a cornice show framing and finish members.

## Basement Wall Detail and Walkout Detail

These details, shown side by side on Sheet 5, are drawn at the scale of ½" = 1'-0". Refer to the Foundation/Basement Plan, Sheet 1 to see the stepped basement wall and to the Plot Plan, Sheet 6 to see the slope of the lot.

The front and sides of the house are 4" brick veneer over a 2 × 6 framed wall. The rear of the house is 12" lap siding over a 2 × 6 framed wall for the portion around the walkout basement. The rear wall below grade is 8" thick concrete. The 6 in 12 sloped roof is framed with 2 × 8 roof rafters spaced 16" OC and 2 × 10 ceiling joists spaced 16" OC. Plywood sheathing is covered with felt and cedar shakes.

Brick veneer walls are framed of 2 × 6s with ½" insulated sheathing applied. A ½" air space separates the face brick from the sheathing. These walls are rated R-19. The rear wall is also framed with 2 × 6s. It is faced with ½" foam board insulation which is covered with the 12" lap siding shown in the elevations. The rear wall is rated R-19.

The foundation footing for the 10" concrete basement wall measures 8" × 20". Three #4 rebars are placed continuously in the foundation footing and #4 rebars are placed horizontally every 2'-0" to reinforce the foundation footing. The concrete wall contains #4 rebars placed horizontally and vertically at 2'-0" OC.

The foundation footing for the rear wall of the house is 8" × 16" with #4 rebars. A 2'-0" foundation wall supports the 4" concrete slab that forms the basement floor. The slab is reinforced with 10 × 10 welded wire fabric. A 6 mil plastic sheet provides vapor protection for the slab.

The main floor of the house is supported by 2 × 10 floor joists spaced 16" OC. Batt insulation is placed 30" in from all exterior walls. A ¾" tongue-and-groove plywood deck is nailed to the floor joists and is covered with carpet and trimmed with baseboard molding.

# Sketching

Name _____ Date _____

## SKETCHING 7-1

**Sketch an elevation showing details of the vanity wall in the space provided. Include the following details.**

1. 4′-0″, two-door vanity, 30″ high

2. 12″ × 12″ × 8′-6″ soffit above vanity

3. 6″ backsplash on vanity

4. 48″ × 48″ mirror above vanity

5. Floor-to-ceiling height of 8′-1″

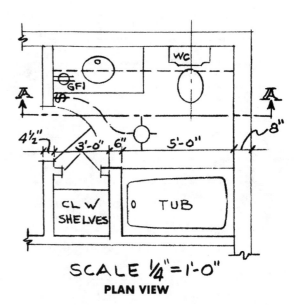

SCALE ¼″=1′-0″

**PLAN VIEW**

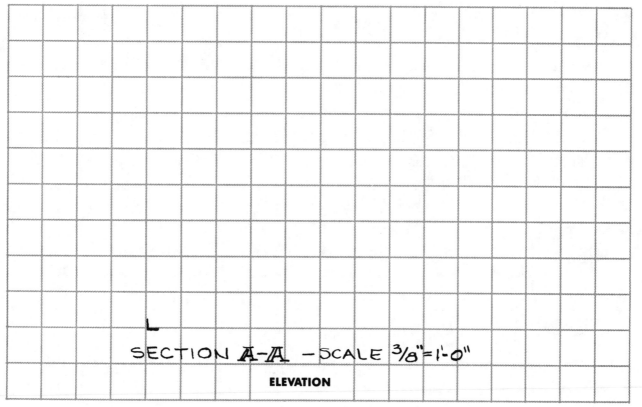

SECTION A-A — SCALE ⅜″=1′-0″

**ELEVATION**

## SKETCHING 7-2

**Sketch full-size details of the moldings in the spaces provided. Refer to the Appendix.**    ¼″

①    ②    ③

½ × ¾ BASE SHOE          ¾ × ¾ COVE          ½ × ½ QUARTER ROUND

④    ⑤

⁷⁄₁₆ × 1⅛ STOP (MODERN)          ⁷⁄₁₆ × 3¼ BASE (COLONIAL)

⑥    ⑦

¹¹⁄₁₆ × 2¼ CASING (MODERN)          ¹¹⁄₁₆ × 2¾ STOOL

⑧    ⑨    ⑩

¼ × 2 MULLION CASING          ¼ × ¾ SCREEN BEAD          ¾ × ¾ CORNER BEAD
(MODERN)          (FLAT)          (ROUND EDGE)

# Review Questions

Name _____ Date _____

## Identification

_____  1. Studs

_____  2. Siding

_____  3. Sash bottom rail

_____  4. Apron

_____  5. Glass light

_____  6. Jamb

_____  7. Inside casing

_____  8. Sheathing

_____  9. Sash top rail

_____ 10. Sash stile

_____ 11. Stop

_____ 12. Outside casing

_____ 13. Meeting rail

_____ 14. Sill

_____ 15. Gypsum board

_____ 16. Drip cap

_____ 17. Stool

_____ 18. Parting strip

_____ 19. Header

_____ 20. Blind stop

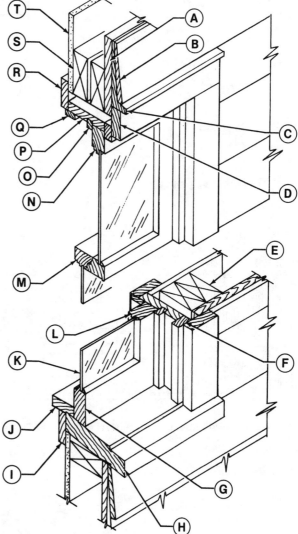

# Details

**Refer to Completion on page 129.**

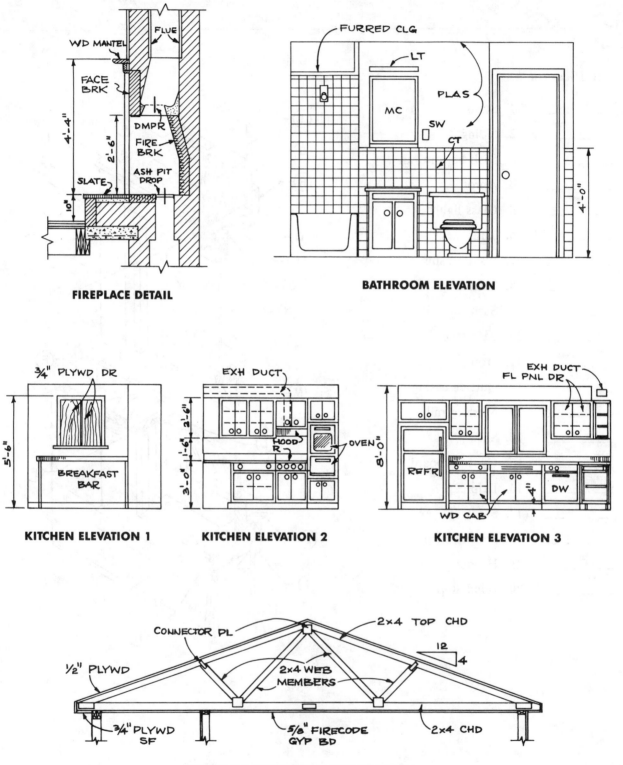

**FIREPLACE DETAIL**

**BATHROOM ELEVATION**

**KITCHEN ELEVATION 1**

**KITCHEN ELEVATION 2**

**KITCHEN ELEVATION 3**

**TRUSSED RAFTER DETAIL FOR GARAGE**

## Completion

**Refer to Details on page 128.**

———————  1. The fireplace hearth is ————" above the finished floor level.

———————  2. The fireplace opening is ———— high.

———————  3. The fireplace hearth is supported by a(n) ———— slab.

———————  4. The fireplace chimney is faced with ———— brick.

———————  5. The fireplace hearth is a piece of ————.

———————  6. The fireplace mantel is made of ————.

———————  7. The fireplace mantel is ———— above the finished floor level.

———————  8. ———— panel doors are used on kitchen cabinets.

———————  9. The number of wall cabinet doors shown is ————.

———————  10. The dishwasher is located to the ———— of the kitchen sink.

———————  11. Pass-through doors from the kitchen to the breakfast bar are ———— above the finished floor level.

———————  12. The kitchen cabinets are made of ————.

———————  13. Wall cabinets are ————" high.

———————  14. Base cabinets are ————" high.

———————  15. The number of base cabinet doors shown is ————.

———————  16. The number of wall ovens shown is ————.

———————  17. A(n) ———— cabinet is placed above the vanity.

———————  18. A(n) ———— ceiling is shown over the tub.

———————  19. ———— tile is 4'-0" above the finished floor level behind the commode and vanity.

———————  20. The garage roof has a(n) ———— in 12 slope.

———————  21. Plywood used to deck the garage roof is ————" thick.

———————  22. The garage soffit is covered with ————" plywood.

———————  23. ———— plates are used to tie chords and web members together.

———————  24. Web members are made of ————.

———————  25. Firecode gypsum board on the garage ceiling is ————" thick.

## True-False

T   F     **1.** Detail views are sectional views, elevation views, or plan views drawn to show additional information.

T   F     **2.** Details are usually drawn to a larger scale than the view from which they are taken.

T   F     **3.** A preferred scale for detail views on residential plans is $\frac{1}{4}'' = 1'\text{-}0''$.

T   F     **4.** Generally, dimensions are not repeated on plans unless necessary for clearness.

T   F     **5.** Size dimensions locate a particular feature in relation to another feature.

T   F     **6.** Arrows on a cutting plane line indicate the direction of sight in which a sectional view is taken.

T   F     **7.** Specific references are required for typical details.

T   F     **8.** The plan view of a kitchen and its elevation views are related.

T   F     **9.** Details of windows are drawn as sectional views through the head, sill, and jamb.

T   F    **10.** Interior details are commonly drawn of cabinets, millwork, and trim.

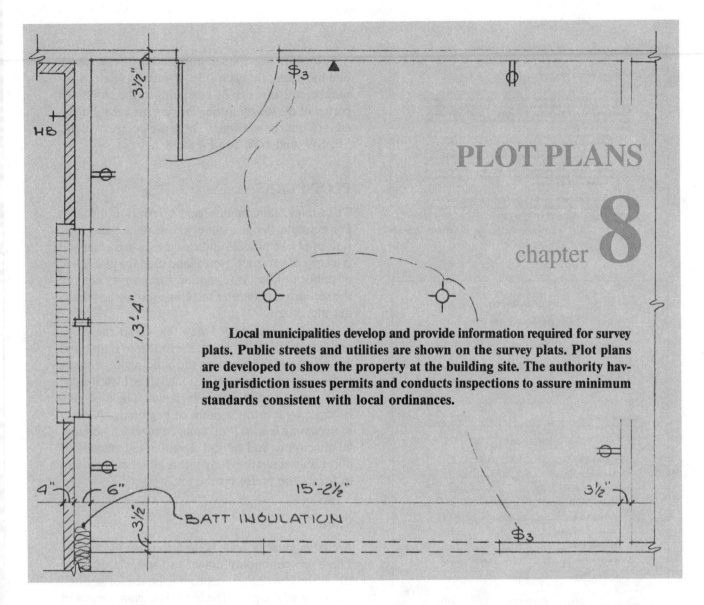

# PLOT PLANS

chapter **8**

**Local municipalities develop and provide information required for survey plats. Public streets and utilities are shown on the survey plats. Plot plans are developed to show the property at the building site. The authority having jurisdiction issues permits and conducts inspections to assure minimum standards consistent with local ordinances.**

## BUILDING CODES

Building Codes are designed to ensure that materials and methods of general construction, plumbing, mechanical, and electrical work used in dwellings meet minimum standards. The authority having jurisdiction in the area enforces local building codes which are commonly based on model codes. For example, the Council of American Building Officials' (CABO) *One and Two Family Dwelling Code* is a model code developed by the Building Officials and Code Administration, Inc. (BOCA), the International Congress of Building Officials (ICBO), and the Southern Building Code Congress International.

Building permits must be obtained after the plans are completed by the architect and before construction begins. The building permit must be displayed on the job site throughout construction. Inspectors enforce the local building codes by checking materials and construction methods during construction. The final inspection is made after the house is completed. The house may be occupied after it passes the final inspection. See Figure 8-1.

## SURVEY PLATS

Basic information for the plot plan is found on the survey plat drawn by a licensed surveyor. A *survey plat* is a map showing land divisions, such as a portion of a quarter section of a township divided into streets and lots. *Townships* are square areas that are six miles long on each side. Townships are subdivided into sections and quarter sections.

**One and Two Family Dwelling Code**

When bearing studs are spaced at 24-inch intervals, and doubled top plates are less than two 2 by 6 or two 3 by 4 members, and supported floor joists, floor trusses or roof trusses are spaced at more than 16-inch intervals, care shall be exercised to ensure locating supported floor joists or trusses within 5 inches of the studs beneath, or, in lieu thereof, solid blocking equal in size to the studs beneath shall be installed to reinforce the double plate above.

**EXCEPTION:** A single top plate may be installed in bearing and exterior walls, provided the plate is adequately tied at joints, corners and intersecting walls by at least the equivalent of 3-inch by 6-inch by 0.036-inch-thick galvanized steel that is nailed to each wall or segment of wall by three 8d nails or equivalent, provided the rafters or joists are centered over the studs with a tolerance of no more than 1 inch. Top plate may be omitted over lintels which are adequately tied to adjacent wall section as previously described or with 1-inch by 4-inch by 12-inch wood members splice nailed to each wall section by three 8d nails or equivalent.

**R-402.4—Interior Partitions.** Interior load-bearing partitions shall be constructed, framed and fire-stopped as specified for exterior walls. Interior nonbearing partitions may be constructed with 2-inch by 3-inch studs spaced 24 inches on center or 2-inch b

*International Congress of Building Officials*

BUILDING CODE

CITY OF JACKSON

**BUILDING PERMIT**

Address ____________________

Lot # ____________________

Type of Work ____________________

General Contractor ____________________

INSPECTIONS

| | |
|---|---|
| Sanitary Lateral ________ | Date ______ |
| Foundation ________ | Date ______ |
| Plumbing Rough ________ | Date ______ |
| Electrical Rough ________ | Date ______ |
| Building Framing ________ | Date ______ |
| Mechanical Systems ________ | Date ______ |
| Plumbing Final ________ | Date ______ |
| Electrical Final ________ | Date ______ |
| Building Final ________ | Date ______ |

Permit Issued ______

Permit Expires ______ ____________ Building Official

BUILDING PERMIT

**Figure 8-1.** Local building codes are based on model codes. Building permits are issued before construction begins.

Townships are based on a gridwork of North-South and East-West lines crisscrossing the United States and forming squares that are 36 miles on each side. North-South lines are *meridians*. East-West lines are *baselines*. Townships are identified by their location in relation to the meridians and baselines. For example, a township three rows North of the baseline and five rows East of the meridian is designated T3N, R5E.

Townships are divided into sections. A *section* is one mile long on each side. Township sections are numbered 1 through 36 beginning at the Northeast corner of the township. Sections are further subdivided into quarter sections. These are designated NE, SE, SW, and NW. See Figure 8-2.

## PLOT PLANS

The survey plat contains legal descriptions of land. For example, the plat survey shows the lot and location of public utilities and easements. An *easement* is a strip of privately owned land used for placement of public utilities. Information from survey plats of the sections and quarter sections is used when drawing plot plans.

A *plot plan* is a scaled view that shows the shape and size of the building lot, the location, shape, and overall size of the house on the building lot, and the finish floor elevation. Additionally, the locations of the street(s) and utilities are shown. The plot plan is begun by locating a point of beginning. A *point of beginning* is a location point from which horizontal dimensions and vertical elevations are made. This point is used to accurately locate all lot corners and to locate the house on the lot.

### Scale

A civil engineer's scale is used to draw plot plans. These are commonly drawn to the scale of 1″ = 10′-0″ or 1″ = 20′-0″ depending on the lot size and sheet size of the plan. Decimal dimensions are used on the plot plan. For example, a property line that is 120′-6″ long is designated 120.5′.

For practical purposes when locating the building on the lot, tradesworkers use tape measures containing feet, inches, and fractions of an inch. Consequently, decimals are converted into fractions. To convert decimals into fractions, divide the decimal by 100 and reduce the fraction. For example, .50 = $\frac{1}{2}$ (.50 divided by 100 equals $\frac{5}{10}$ which is reduced to $\frac{1}{2}$). To convert fractions into decimals, divide the *numerator* (number above the fraction bar) by the *denominator* (number below the fraction bar). For example, $\frac{3}{8}$ = .375 (3 divided by 8 equals .375).

### Elevations

*Elevations* are vertical measurements above or below the point of beginning. (*Note:* The use of the word

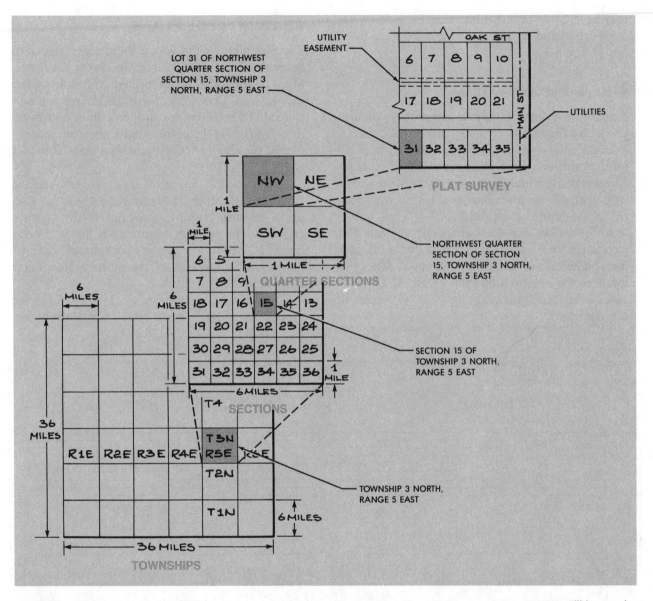

**Figure 8-2.** Townships are subdivided into sections and quarter sections. The plat survey shows streets, utilities, and easements.

elevation in this sense is different from the elevation view of a house.) The point of beginning is shown on the plot plan. A bench mark on the actual lot represents the point of beginning. A *bench mark* is a stake driven into the ground or a point in the street or curb from which all measurements are begun. For example, a bench mark with an elevation of 112′ shows that point to be 112′ above the point of reference used by the community. The point of reference for a community may be in relation to sea level or to a specific point in the community.

**Contour Lines.** *Contour lines* are dashed or solid lines on a plot plan drawn to pass through points having the same elevation. Dashed lines represent the natural grade. Solid lines represent the finish grade. *Natural grade* is the slope of the land before rough grading. *Finish grade* is the slope of the land after final grading.

Closely spaced contour lines denote steeply sloped lots. Contour lines spaced farther apart denote flatter lots. The spacing of contour lines is generally at 1′-0″ intervals. For steeply sloped lots, the spacing may be at 2'-0" or larger intervals. Elevation nota-tions are placed on the high side of contour lines. See Figure 8-3.

## Symbols and Abbreviations

Symbols and abbreviations conserve space on plot plans. See Figure 8-4. Symbols show roads, streets, sidewalks, the point of beginning, property lines, all utilities, the natural and finish grades, and other in-formation. Trees and shrubbery to remain are shown with symbols. North is indicated with a symbol to orient the house on the lot.

Abbreviations on plot plans relate to roads and streets, the lot, location of the house on the lot, util-ities, building materials, elevations, and so forth. For example, the abbreviation FIN. FL EL 32.25' indi-cates a finished floor elevation of 32'-3". Refer to the Appendix.

## READING PLOT PLANS

The plot plan for the Wayne Residence is drawn to the scale of 1" = 20'-0". Refer to Wayne Residence, Sheet 6. The lot, which is located on Country Club Fairways in Columbia, Missouri, is pentagonal-shaped. Dimensions for property lines and nota-tions indicating the corner angles are given. For ex-ample, the angle of the Southeast corner of the lot is 104°-23'-25".

A 9'-0" utility easement is shown between the curb and the property line along the street. Another utility easement is shown along the Northwest prop-erty line. The point of beginning is the Southwest corner of the lot. The corner of the garage is 25'-0" North and 11'-0" East of the property line. Refer to the Floor Plan, Sheet 2.

The finish floor elevation is 725'-0". Refer to the Floor Plan, Sheet 2 for the overall size of the house.

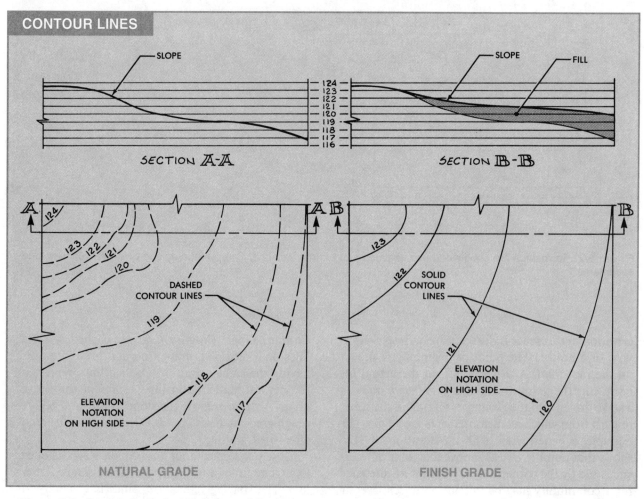

**Figure 8-3.** Contour lines show the slope of the lot.

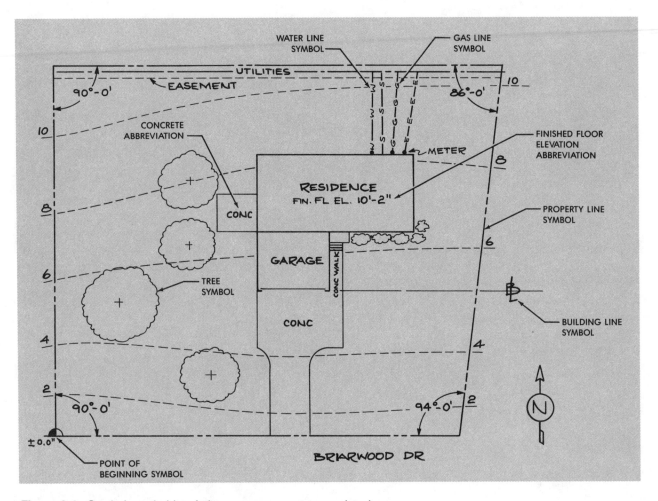

**Figure 8-4.** Symbols and abbreviations conserve space on plot plans.

A concrete drive and 4'-0" concrete walk are shown at the front of the house.

Contour lines show the finish grade of the lot. Numbers on the contour lines give elevations. The lot slopes down from West to East. The elevation near the Southwest corner of the house is almost 726' while the elevation at the Southeast corner of the house is 718'. Retaining walls are shown on the West and North sides of the house. Refer to the Foundation/Basement Plan, Sheet 1, and the North and East Elevations, Sheet 3 for additional information regarding the retaining walls. The closely spaced contour lines from the retaining wall on the North side of the house represent a steep slope in the finish grade.

Symbols are used to show four trees that are to remain on the lot. Additionally, symbols show foundation shrubbery along the front of the house between the front entrance and the garage.

# Sketching

Name _____ Date _____

## Sketching 8-1

**Sketch symbols for plot plans as indicated. Refer to the Appendix.**

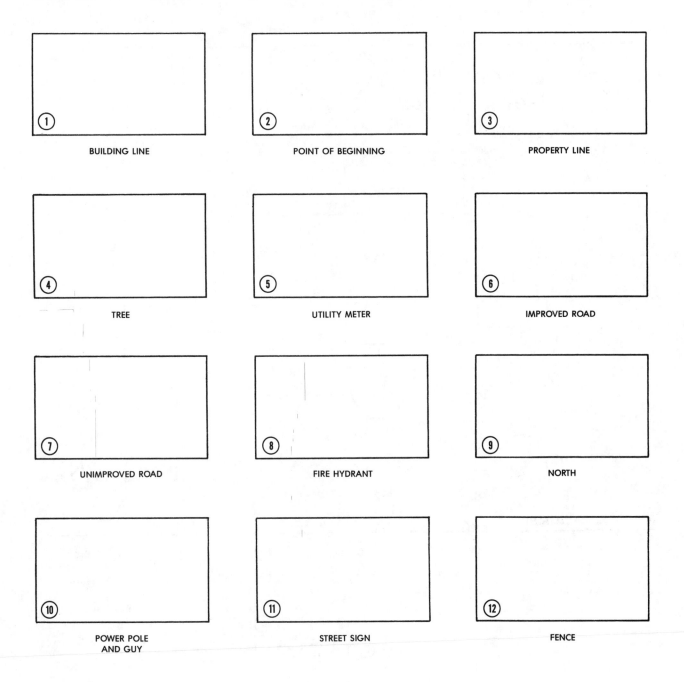

| | | |
|---|---|---|
| ① | ② | ③ |
| BUILDING LINE | POINT OF BEGINNING | PROPERTY LINE |
| ④ | ⑤ | ⑥ |
| TREE | UTILITY METER | IMPROVED ROAD |
| ⑦ | ⑧ | ⑨ |
| UNIMPROVED ROAD | FIRE HYDRANT | NORTH |
| ⑩ | ⑪ | ⑫ |
| POWER POLE AND GUY | STREET SIGN | FENCE |

## Sketching 8-2

**Sketch and/or letter the following to complete the plot plan.**

1. 512' (elevation notation)
2. Existing trees (symbol)
3. 150.5' (dimension)
4. STORM DRAIN (symbol and notation)
5. WATER MAIN (symbol and notation)

6. GAS MAIN (symbol and notation)
7. SANITARY SEWER (symbol and notation)
8. POINT OF BEGINNING (notation)
9. 12.5' (dimension)
10. FIRST FL ELEV 513.5' (notation)

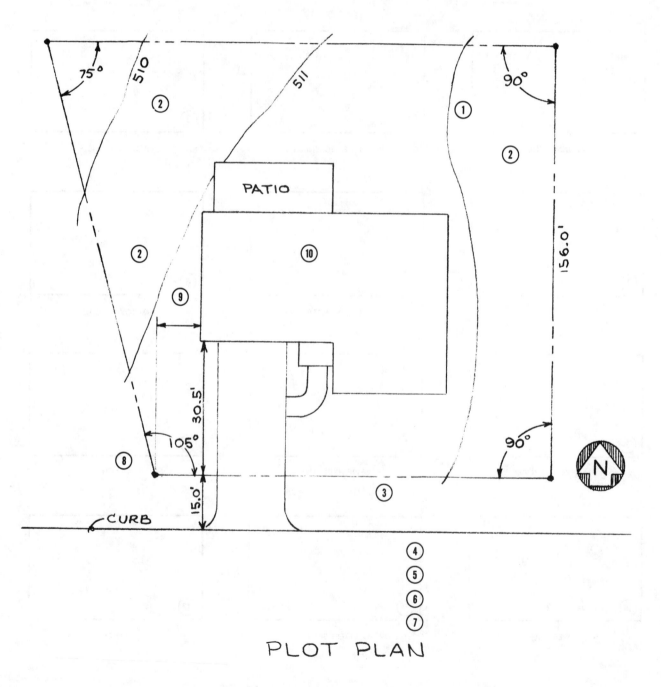

PLOT PLAN

# Review Questions

Name _____ Date _____

## True-False

T    F    **1.** Public streets and utilities are shown on survey plots.

T    F    **2.** The authority having jurisdiction in an area inspects the house during construction.

T    F    **3.** Building permits must be obtained before the plans are completed.

T    F    **4.** Townships are square areas one mile long on each side.

T    F    **5.** Baselines on maps showing townships run North and South.

T    F    **6.** A section contains one square mile.

T    F    **7.** Easements are private property containing public utilities.

T    F    **8.** Property dimensions on plot plans are generally given in feet and decimal parts of a foot.

T    F    **9.** The decimal .675 equals ⅝.

T    F    **10.** Solid contour lines show the natural grade of the lot.

T    F    **11.** Closely spaced contour lines on a plot plan represent a steep slope.

T    F    **12.** Trees and shrubbery should not be shown on plot plans.

T    F    **13.** The symbol showing North should always be included on a plot plan.

T    F    **14.** Utilities on plot plans may be shown with symbols and notations.

T    F    **15.** Spacing of contour lines for plot plans is generally at $1'-0''$ intervals.

## Multiple Choice

_____    **1.** The _____ scale is used to draw plot plans.
          A. architect's
          B. civil engineer's
          C. mechanical engineer's
          D. tradesworker's

_____    **2.** A point of beginning is the location point for _____ dimensions.
          A. horizontal
          B. vertical
          C. both A and B
          D. neither A nor B

_____ 3. A township four rows North of the baseline and two rows East of the meridian is designated _____.
  A. T2N, R2E
  B. R3E, T3N
  C. T2N, R3N
  D. none of the above

_____ 4. Townships are divided into _____.
  A. quarter sections
  B. communities
  C. towns and farms
  D. none of the above

_____ 5. A plot plan shows the _____.
  A. size of the building lot
  B. public streets adjacent to the building lot
  C. finish floor elevation
  D. all of the above

_____ 6. A four-sided building lot with two 90° corners and one 83° corner also contains one _____° corner.
  A. 89
  B. 92
  C. 97
  D. 107

_____ 7. A plot plan showing a finish floor elevation of 132'-10" is _____ above the point of beginning at an elevation of 128.5'.
  A. 3'-8"
  B. 4'-4"
  C. 4'-5"
  D. 4'-8"

_____ 8. A rectangular building lot measuring 126'-0" × 212'-0" contains _____ square feet.
  A. 252
  B. 338
  C. 22,424
  D. 26,712

_____ 9. The numerator of a fraction is the number _____ the fraction bar.
  A. above
  B. below
  C. either A or B
  D. neither A nor B

_____ 10. Elevations may be viewed in relation to the plot plan to obtain additional information concerning the _____ of the house.
  A. room layout
  B. room size
  C. interior
  D. exterior

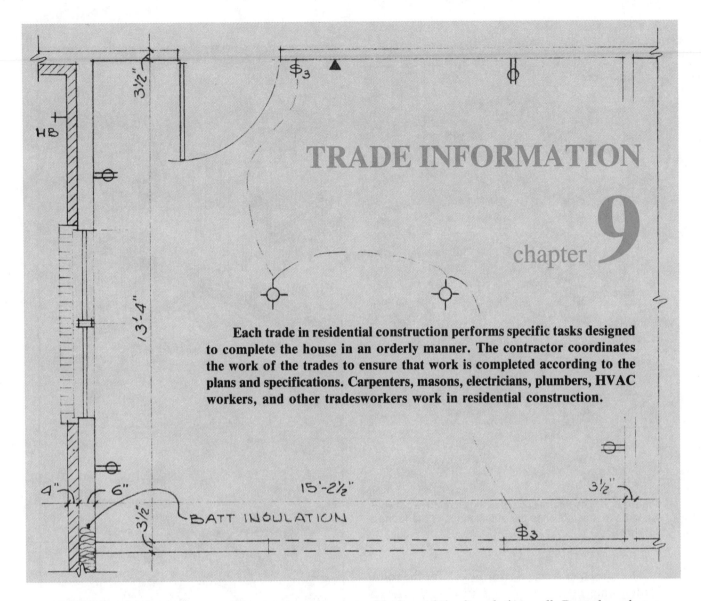

## TRADE INFORMATION

chapter **9**

Each trade in residential construction performs specific tasks designed to complete the house in an orderly manner. The contractor coordinates the work of the trades to ensure that work is completed according to the plans and specifications. Carpenters, masons, electricians, plumbers, HVAC workers, and other tradesworkers work in residential construction.

## CARPENTRY

Carpenters erect forms for foundation footings and walls, concrete walks, driveways, and basement and garage floors. They install floors, frame walls and roofs, and apply exterior and interior trim. Information required by carpenters is found in the specifications, on plot plans, floor plans, sectional views, details, elevations, and on schedules.

## Concrete Foundation Work

Concrete foundation work begins with the establishment of the corners of the building in relation to the point of beginning, which is shown on the plot plan. Stakes are set up to give the excavator the size and depth of the excavation. Batterboards are erected and strings are stretched to indicate the out-

side face of the foundation wall. Batterboards are set back from the excavation and are used for checking purposes at several steps in the construction of the footings and erection of the foundation forms. See Figure 9-1.

Footings are usually poured first as a separate operation. The formwork for footings is placed so that the foundation wall is centered on the footings, and the top of the footings are level and at the propper height. Rebars are placed in the footings during the concrete pour to increase the strength of the concrete. *Rebars* are steel rods used for reinforcing concrete structural members. Refer to the Appendix. After the footings are poured, a keyway is formed in the footings. A *keyway* is a groove in one lift of concrete that is filled with concrete of the next lift. A *lift* is the amount of concrete placed in one pour.

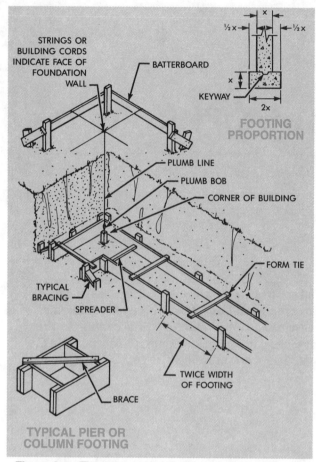

**Figure 9-1.** The intersection of strings stretched between batterboards establishes the outside face of the foundation wall.

The carpenter erects the formwork for the foundation walls after the footings are poured. Generally, foundation walls for residential construction are 8″ thick and high enough to provide adequate headroom in the basement. Two common types of forms used in residential construction are built-in-place forms and panel forms. Form ties with steel wedges or snap ties with cones may be used to maintain wall thickness during the concrete pour. See Figure 9-2.

Built-in-place forms are constructed on the job site. The *sole* (bottom plate) is fastened to the concrete footing. Studs are set up, temporarily tied together, and braced. Walers, either single or double, are secured to the outsides of the studs. *Walers* are horizontal pieces placed on the outsides of form walls to strengthen and stiffen the walls. Sheathing is then applied to the inner faces of the studs.

*Panel forms* are sections made of studs and plates nailed to a plywood panel. The end studs are fas-

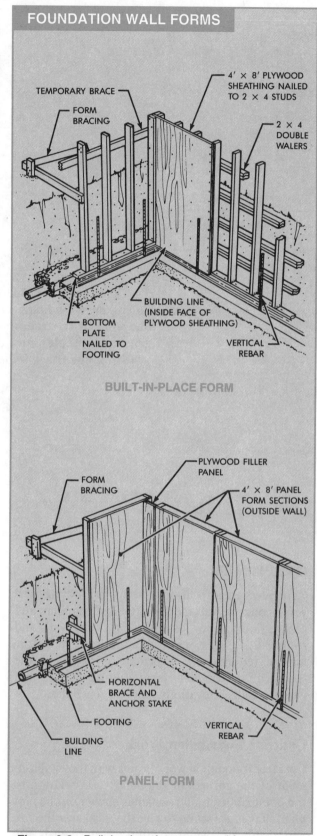

**Figure 9-2.** Built-in-place forms or panel forms are used to form foundation walls.

tened together with duplex nails to form the foundation wall. *Duplex nails* have two heads that facilitate stripping of forms. Panel form sections may be built on the job or off site. They are especially convenient when one housing design is repeated for several houses. Patented panel systems of wood or metal may be rented or purchased.

Slab-on-grade foundations are commonly used in warmer climates where the frost line presents no heaving and shifting problems. A *slab-on-grade foundation* is a ground-supported foundation system consisting of shortened foundation walls, or a thickened edge, and a concrete slab. Such systems used in colder climates require edge insulation. Slabs for residential construction are commonly 4″ to 8″ thick and may be thickened for heavier loads. See Figure 9-3.

## Walks, Driveways, and Floors

Carpenters erect the forms for walks, driveways, and garage floors. This is normally done well into the construction of the dwelling so that the concrete has time to set and cure without being disturbed. Driveways and garage floors are often reinforced with welded wire fabric to help prevent cracking. See the Appendix. After finishing, the surface is kept damp for several days as the concrete cures. Forms are removed after the second day.

## Floors, Walls, and Roofs

Floor construction begins after the foundation footings and walls are poured and the forms are stripped. Carpenters refer to the floor plans and details of sectional views through exterior walls to determine materials to be used and dimensions. Sill plates are attached to the top of the foundation wall and header joists and regular joists are nailed in place. A notation on the floor plan gives the size and direction of floor joists.

Floor joists commonly run across the full width of the house and are supported at the ends on a sill plate resting on the foundation wall. Header joists maintain alignment of the floor joists. Longer joists require additional support. These joists may be supported by a girder or steel beam.

*Girders* in residential construction are commonly laminated (built-up) structural members designed to carry heavy loads. Floor joists should either butt together or be lapped over the girder or steel beam.

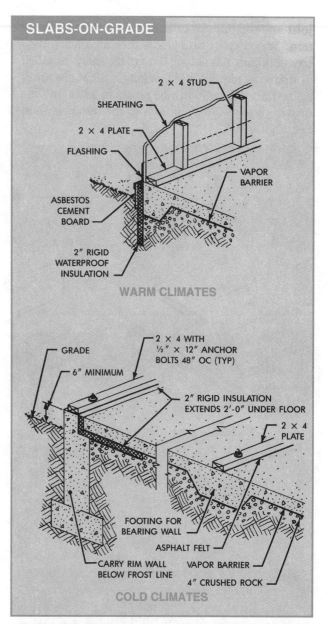

**Figure 9-3.** Slab-on-grade foundations are common in warm climates. They must be insulated for use in cold climates.

Butted joists are joined by wood or metal ties. Steel posts may be required for intermediate support of girders and beams depending on their length, size, and weight to be supported.

Bridging is often placed between floor joists. *Bridging* is bracing between joists or studs that adds stiffness to the floors or walls. The two common types of bridging are cross bridging and solid bridging. Cross bridging may be wood (commonly 1″ × 3″ or 2″ × 3″) or metal attached to the top of one

ad the bottom of the next joist in an X pat-
Wood cross bridging is toenailed and metal
cross bridging has nailing flanges that may be nailed
or driven in place. Solid bridging is the same width
as the floor joists. It may be placed in a straight line
or staggered, which is preferred for faster nailing.
See Figure 9-4.

After floor joists are in place, the floor is placed
and walls are framed. Carpenters refer to the floor
plans and wall details for dimensions. The length
and height of walls, size and spacing of studs, and
location and size of rough openings for doors and
windows are determined. Framed walls are commonly built-up on the floor and raised into posi-

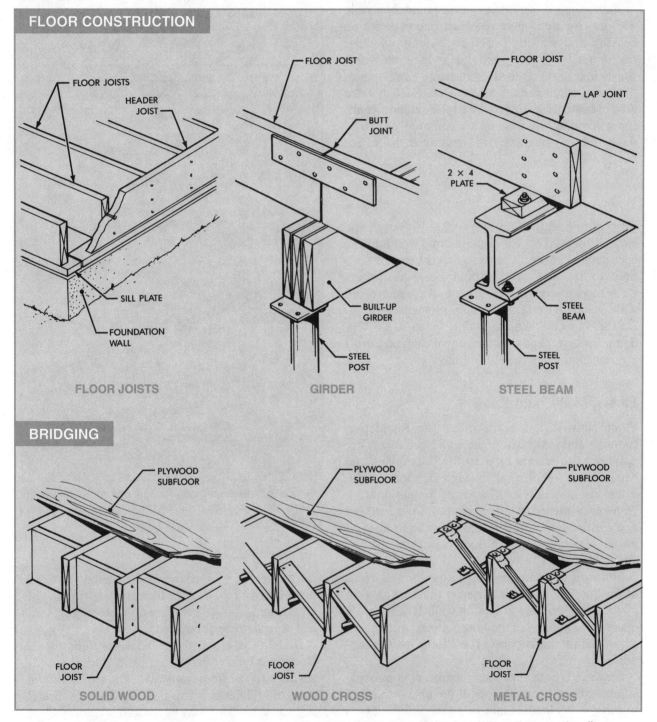

**Figure 9-4.** Notations on floor plans give size and direction of floor joists. Bridging stiffens floors and walls.

tion. They are then plumbed, squared, and secured to the framing members of the floor. Room layout, particularly for baths and kitchens, is critical, and frequent reference to the floor plans is required. See Figure 9-5.

Roofs are framed after the walls are in place. Sectional views and details of wall and roof framing indicate material type and size and roof overhang. Notations on floor plans give the size and direction of ceiling joists. Roof framing plans are often included in the set of plans to provide comprehensive roof framing information. Elevations give the slope of the roof and show finish materials to be applied. Roofs may be constructed of rafters or trusses. In rafter construction, ceiling joists and rafters are built up on the job site. In truss construction, trusses are commonly delivered to the job site and installed. See Figure 9-6.

## Exterior and Interior Trim

The specifications and plans provide information needed to finish the exterior and interior of the building. Finish items detailed in the specifications include door and window hardware, bath fixtures, electrical fixtures, appliances, wall and floor materials and finishes, and so forth. Carpenters fit and apply all trim such as window and door casings and exterior and interior molding. Refer to the Appendix.

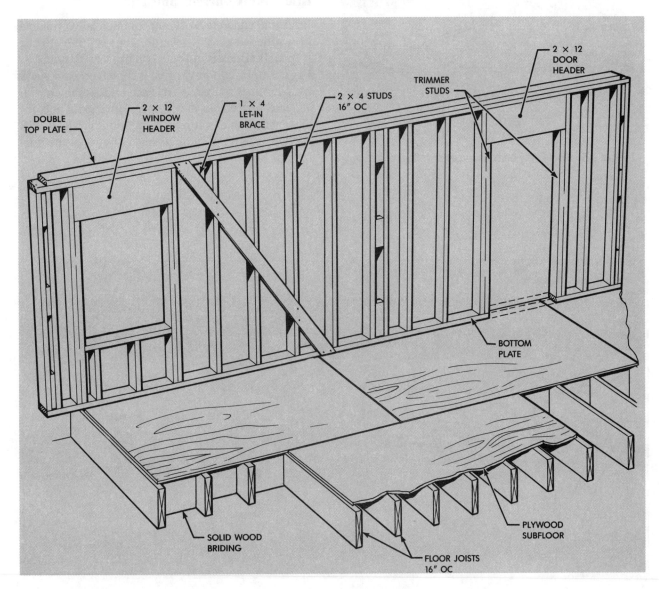

**Figure 9-5.** Floor plans and details give dimensions for framing the walls.

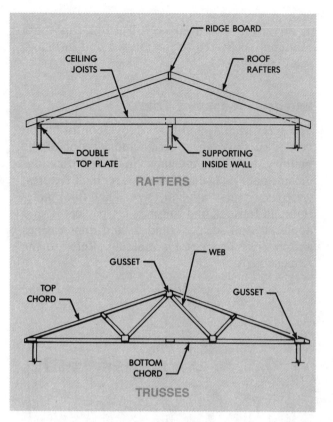

**Figure 9-6.** Roofs may be constructed of rafters or trusses.

## MASONRY

Masonry work includes the laying of brick, concrete block, tile, and stone. Elevations show the type of brick to be used and the walls to be bricked. Sectional views (often drawn as details) through the exterior walls show how bricks are to be laid in the wall at windows and doors, and how the top of the wall is to be finished at the eaves. Floor plans give dimensions showing the exact location of openings for windows and doors. Masonry walls include brick, block, brick and block, and brick veneer walls. See Figure 9-7.

### Brick and Concrete Block

Brick and concrete block are commercially available in a wide range of sizes and shapes. Refer to the Appendix. Brick is indicated on prints with symbols. For example, common brick is shown on plan views with 45° lines spaced apart. Face brick is shown with 45° lines spaced closer together. Horizontal lines indicate brick on elevations. A notation such as FACE BRICK is lettered in open spaces between the horizontal lines.

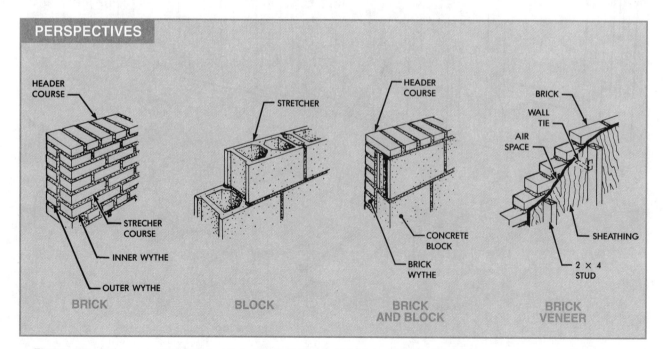

**Figure 9-7.** Masonry walls include brick, block, brick and block, and brick veneer.

Brick may be either standard or modular in size. *Standard brick* is classified by its nominal size. For example, a face brick is nominally 2″ × 4″ × 8″, but its actual size is approximately 2¼″ × 3¾″ × 8″. *Modular brick*, designed so that every third horizontal joint falls on a multiple of 4″, is classified by its actual size. Brick is laid in various positions and bonds as noted on the plans. *Brick bond* is the pattern formed by the exposed faces of the brick.

Concrete block is shown in elevation views with horizontal and vertical lines and a series of dots and small circles or triangles. Standard concrete block is 7⅝″ × 7⅝″ × 15⅝″. When laid with ⅜″ mortar joints, it measures 8″ × 8″ × 16″ and fits the 4″ module of modular measure.

## ELECTRICAL

Supply conductors from the utility company to the premises are regulated by the *National Electrical Safety Code* (NESC). Service-entrance conductors and electrical work from the meter base to the service equipment and the dwelling are governed by the current National Electrical Code® (NEC®). The purpose of the NEC® is the practical safeguarding of persons and property from hazards that may arise from using electricity. The NEC® is published by the National Fire Protection Association.

The NEC® is advisory. It is adopted by states, counties, and municipalities. The NEC® is updated every three years. In addition to the NEC®, local ordinances are enforced. Always check with the authority having jurisdiction in the area where electrical work is performed.

### Service

Electrical service is grounded at the service equipment to provide a common grounding terminal for the grounded neutral, equipment grounding conductors, and the grounding electrode conductor. A main bonding jumper grounds the terminal bar to the metal enclosure connected to the grounding electrode conductor. Feeders and branch circuits are routed from the service equipment to subpanels and loads.

Electrical service to a dwelling unit is provided by an *overhead drop* or a *lateral* (underground). Clearances for overhead drops are regulated by 230-24(b) of the NEC® and Table 232-1 of the NESC. Burial depths for laterals are regulated by

Table 300-5. Local ordinances may amend these clearances and depths. The power line is shown on the plot plan by a line consisting of a series of intermittent long and short dashes. See Figure 9-8.

### Conductors

Copper (Cu) and/or aluminum (Al) conductors are used to wire dwellings. Copper conductors are more expensive than aluminum conductors. Copper conductors also have less resistance (for the same size) to restrict current flow from the overcurrent protection device to the load. Aluminum conductors in larger sizes are commonly used for service-entrance conductors and feeders to supply power to subpanels. Table 310-16 gives conductor sizes and ampacities for copper and aluminum or copper-clad aluminum conductors according to temperature ratings.

**Service-entrance Conductors.** Service-entrance conductors shall have at least a 100 A rating where the calculated load is 10 kVA or greater. A 100 A service shall be provided if there are six or more 2-wire branch circuits. The minimum size for service-entrance conductors for any dwelling is #6.

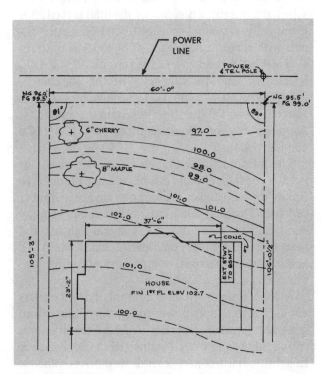

**Figure 9-8.** The location of power lines is shown on the plot plan.

Service-entrance conductors are routed in a raceway or service-entrance cable. Insulation is TW, THW, THWN, XHHW, RHW, or other types listed for wet locations. See Table 310-13 for conditions of use for each type of insulation.

Service-entrance conductors are sized from the total VA rating of the dwelling unit. The formula $I = VA/V$ is used to find amps. In this formula $I$ = amps, $VA$ = volt-amps, and $V$ = volts. The letter $A$ is also used to designate amps. Volt-amps are divided by volts to obtain ampacity ($I = VA/V$). For example, a calculated load of 36,000 VA on a 240 V, 1PH service has an ampacity of 150 A (36,000 VA/240 V = 150 A). Per Table 310-16, #1/0 THWN Cu conductors are required.

**Feeder and Branch Circuit Conductors.** Feeder circuits include all conductors between the service equipment and subpanels. Branch circuits include all conductors from the last overcurrent protection device to the load.

The minimum size feeder circuit conductor is #10 per 215-2. Feeder circuits can be any size required to supply a subpanel or taps to individual loads. Wiring methods for feeder circuits include conductors in raceways, service-entrance (SE) cable, or metal-clad (MC) cable. Feeder circuit conductors are selected from the VA rating of the load served. For example, a subpanel on a 240 V, 1PH supply circuit with a connected load of 20,400 VA requires #4 THWN Cu conductors. The formula, $I = VA/V$, is used to find the ampacity of the conductors (20,400 VA/240 V = 85 A). The conductors are then selected from Table 310-16 based on the ampacity. An ampacity of 85 A requires #4 THWN Cu conductors.

The minimum size branch circuit for dwellings is 15 A. Branch circuits can be any size required to supply the load and can supply one load or a number of loads. Typical branch circuits in dwellings supply combination or single loads. General lighting and receptacle outlets are combination loads. An A/C, heating unit, disposal, compactor, or similar appliance is a single load.

A 15 A branch circuit commonly has 10 outlets connected to supply power to lighting fixtures and general-purpose receptacles. A 20 A branch circuit commonly has 13 outlets connected to supply power to lighting fixtures and general-purpose receptacles. The minimum VA per outlet is 180 VA per 220-3(c)(5). Total outlets permitted is determined by

dividing 180 VA by 120 V to obtain 1.5 A, then dividing 1.5 A into the rating of the overcurrent protection device. See Figure 9-9.

## Raceway Systems

*Raceways* are enclosed channels designed to protect electrical conductors. They may be metal or of insulating material. The type(s) of raceway system used for a specific dwelling is given in the specifications for plans and governed by codes adopted for local use. Many cities permit nonmetallic-sheathed cable (commonly called Romex) to be used for wiring dwelling units. Other cities, Chicago for example, require that all conductors in residential construction be routed in metal conduit.

The most common raceway systems used in residential construction are electrical metallic tubing (EMT), commonly know as thinwall, electrical nonmetallic tubing (ENT), rigid nonmetallic conduit (PVC), and multiple-conductor cable assemblies such as AC armored cable (BX) and nonmetallic-sheathed cable (NM and NMC) commonly known as Romex. Additionally, service-entrance cable (SE, ASE, and USE) are used in applications requiring protection from severe mechanical abuse. See Figure 9-10.

**Electrical Metallic Tubing (EMT).** Electrical metallic tubing, known in the trade as thinwall, is similar to rigid metal conduit but is about 40% lighter. It is used in both exposed and concealed work. Thinwall may be used in most locations, but because of its light weight, thin wall, and use of either compression or set screw couplings and connectors, it cannot be used where subjected to severe physical damage. It is commercially available in 10 ' lengths from $1/2''$ to 4 " in diameter. The $1/2''$ and $3/4''$ sizes are commonly used in residential work.

EMT must be supported every 10 ' and within 3 ' of each outlet box, junction box, cabinet, or fitting. The total number of degrees in all bends in any run shall not exceed 360°. Studs and joists may be notched or drilled to accommodate EMT. Notches, which are preferred, should be as narrow as possible and no deeper than necessary to minimize weakening of the structural members.

**Electrical Nonmetallic Tubing (ENT).** Electrical nonmetallic tubing is a flexible, corrugated raceway made of plastic. Trade sizes of $1/2''$ to 1 " diameters

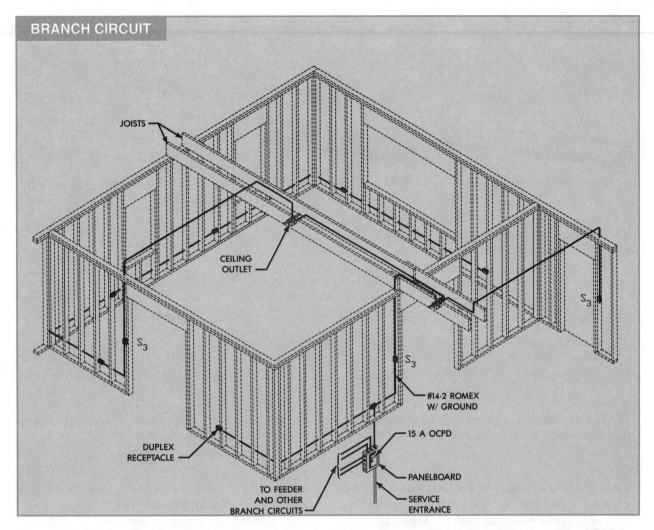

**Figure 9-9.** The number of outlets permitted on a branch circuit is determined by dividing the rating of the OCPD by 1.5 A. The NEC® does not limit the number of outlets.

may be installed in walls, floors, ceilings, and above suspended ceilings having a 15-minute fire rating. ENT may be bent by hand to a radius of not less than 4″ for ¹/₂″ size and not less than 5³/₄″ for 1″ size. The total number of degrees in all bends in any run shall not exceed 360°.

ENT must be supported every 3′ and within 3′ of each outlet box, junction box, cabinet, or fitting. All cut ends must be trimmed and bushings or adapters must be used to prevent abrasion.

**Rigid Nonmetallic Conduit (PVC).** Rigid nonmetallic conduit, known in the trade as PVC (polyvinyl chloride), is waterproof, rustproof, and rotproof. Types of rigid nonmetallic conduit are Schedule 40 and Schedule 80. Schedule 40 (thinwall) does not have the strength of metal conduits and should not

be used where it may be subjected to physical damage. Schedule 80 (heavy wall) may be installed in almost any location where rigid metal conduit may be used.

PVC can be cut with a hacksaw and glued together with couplings and connectors. The total number of degrees in all bends in any run shall not exceed 360°. All field bends must be made with bending equipment. Supports shall be provided within 3′ of each box, cabinet, or connection. The maximum spacing between supports is based upon the diameter of the PVC. For example, the maximum spacing in runs of ¹/₂″ to 1″ diameter PVC is 3′.

**AC Armored Cable (BX).** AC armored cable, known in the trade as BX, is a fabricated assembly of insulated conductors in a flexible metallic en-

## RACEWAY SYSTEMS

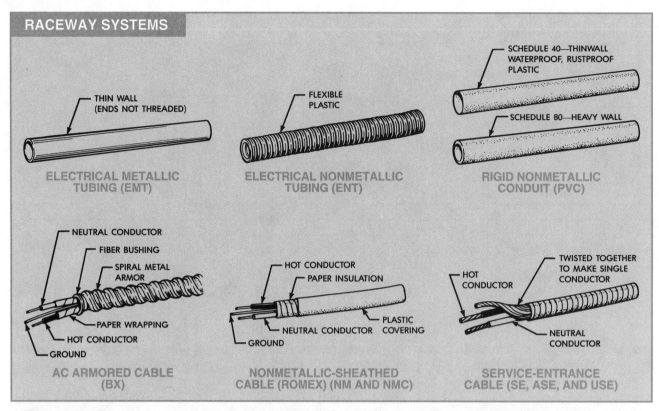

THIN WALL
(ENDS NOT THREADED)

ELECTRICAL METALLIC
TUBING (EMT)

FLEXIBLE
PLASTIC

ELECTRICAL NONMETALLIC
TUBING (ENT)

SCHEDULE 40—THINWALL
WATERPROOF, RUSTPROOF
PLASTIC

SCHEDULE 80—HEAVY WALL

RIGID NONMETALLIC
CONDUIT (PVC)

NEUTRAL CONDUCTOR
FIBER BUSHING
SPIRAL METAL
ARMOR
PAPER WRAPPING
HOT CONDUCTOR
GROUND

AC ARMORED CABLE
(BX)

HOT CONDUCTOR
PAPER INSULATION
PLASTIC
COVERING
NEUTRAL CONDUCTOR
GROUND

NONMETALLIC-SHEATHED
CABLE (ROMEX) (NM AND NMC)

HOT
CONDUCTOR
TWISTED TOGETHER
TO MAKE SINGLE
CONDUCTOR
NEUTRAL
CONDUCTOR

SERVICE-ENTRANCE
CABLE (SE, ASE, AND USE)

**Figure 9-10.** Raceway systems for one-family dwellings include electrical metallic tubing, electrical nonmetallic tubing, rigid nonmetallic conduit, AC armored cable, nonmetallic-sheathed cable, and service-entrance cable.

closure. It may be used in dry locations where not exposed to physical damage. An integral grounding conductor assures a good path for fault current to return from the point of fault to the grounded conductor connected to the grounded busbar in the main service panel.

BX shall be supported every $4\frac{1}{2}'$ and within $12''$ of outlets and fittings. It may be run through holes drilled in studs or through notches in the studs. Holes shall be at least $1\frac{1}{4}''$ from the edge of the stud. Notches shall be covered with $\frac{1}{16}''$ thick steel plates. Bends are permitted but shall be no greater than five times the diameter of the cable when measured along the inside edge. For example, the maximum bending radius of $\frac{1}{2}''$ cable shall not exceed $2\frac{1}{2}''$.

**Nonmetallic-sheathed Cable (NM and NMC).** Nonmetallic-sheathed cable, known in the trade as Romex, has two, three, or four conductors with a green insulated or bare grounding conductor. It is used for economical wiring of dwellings and small commercial buildings. Romex is installed with staples or connectors and must be supported every $4\frac{1}{2}'$ and

within $12''$ of outlets and fittings. Metal or nonmetallic boxes may be used with Romex.

NM cable has a flame-retardant and moisture-resistant outer jacket. It is restricted to inside wiring. NMC cable also has a flame-retardant and moisture-resistant outer jacket. It can be used outside. Neither type shall be buried in concrete.

Romex may be run through the centers of drilled studs, joists, and rafters. Additionally, a notch may be cut in the edge of a stud and covered by a $\frac{1}{16}''$ thick steel plate.

**Service-entrance Cable (SE, ASE, and USE).** Service-entrance cable is used for service-entrance wiring or general interior wiring. The neutral may be insulated or bare when used for service-entrance wiring. An insulated neutral is required when used for general interior wiring. SE cable has a flame-retardant, moisture-resistant outer covering. It is unarmored and will not withstand severe mechanical abuse. ASE cable has an armored coating for additional protection. USE cable is moisture-resistant with an unarmored coating for underground use.

Cables may be routed through studs, joists, and rafters. The bored hole in wooden members shall not be closer than $1^1/_4$" to the edge. Notched members shall be covered with a $^1/_{16}$" thick steel plate. Bends in unarmored cable shall not have a radius less than five times the diameter of the cable. The cable shall be supported at intervals not exceeding $4^1/_2$' and within 12" of outlets and fittings.

## Plans and Schedules

Specifications describe the material and fixtures to be used when wiring a house. Trade names and catalog numbers are often given. Plans show where lighting outlets and receptacles are to be located and where they are switched. Branch-circuit schedules provide information for lighting outlets, switches, and receptacles. Lighting fixture schedules provide information regarding the type and location of lighting fixtures required.

Dimensions are not given to locate receptacles and outlets shown on the floor plan. Long dashed lines (representing circuits) on the floor plan connect receptacles and outlets and switches to show how the outlets are controlled, not specifically where they are located. The electrician determines the exact location and installs them per 210-52. No point along the floor line in any wall 2'-0" or more in length shall be more than 6'-0" from a receptacle in that wall space. Corners unbroken at the floor line are included. Receptacles, in general, should be spaced equal distances apart.

In kitchens, receptacles are installed in each countertop space of 12" or more. Receptacles for fastened-in-place appliances, such as disposals, or receptacles for appliances occupying dedicated space, such as refrigerators, are not included. In bathrooms, one wall receptacle is installed adjacent to the basin. Additionally, one receptacle each is required for the laundry area, basement, garage, and exterior of the house. All 15 A and 20 A, 125 V receptacles in bathrooms, garages, outdoors (with grade level access), and in kitchens within 6'-0" of the sink shall have GFI protection per 210-8(a). Additionally, at least one basement receptacle shall have GFI protection.

Branch circuit schedules and lighting fixture schedules supplement information shown on the floor plan. See Figure 9-11. The service drop is located on the Northwest corner of the house. Service-entrance conductors are run in rigid conduit to the meter and then to the main power panel. All branch circuits are protected by fuses or circuit breakers. Circuits 1 through 8 are shown. Circuit 1 includes all receptacles and lighting outlets for BR 1 and BR 2. The arrow indicates a home run to the main power panel. In BR 1, one type A ceiling outlet is controlled by a single-pole wall switch. Four wall receptacles are installed per 210-52. In BR 2, one type A ceiling outlet is controlled by a single-pole wall switch and four wall receptacles are shown.

## PLUMBING

Plumbing is installed according to information provided on plan views, specifications, and piping drawings. The plumber uses this information and applicable building codes and ordinances to determine installation requirements and procedures. House design and structural configuration are considered when installing plumbing lines. For example, the direction of joists and the location of plumbing fixtures have a bearing on the location of pipes.

Plumbing installation occurs throughout many stages of the construction process. Connections to the water supply, the sanitary sewer, and storm water drainage systems are completed in the early stages following excavation. Basement drains are placed before the basement slab is poured. Houses with a slab-on-grade floor require supply water and waste water piping installed before concrete is placed. The majority of plumbing work is completed after rough framing is completed. This allows access to the inside of walls for installation of supply, waste water, and vent piping. Installation of plumbing fixtures is completed after the walls are covered.

### Piping

The major components of a plumbing system are supply water piping, waste piping, drainage piping, and vent piping. See Figure 9-12. In some regions, the plumber is responsible for installing gas piping to the house from the main supply as specified by local codes and job practices.

**Supply Water Piping.** Supply water piping conveys potable water to the points of use in the house. *Potable water* is water free of impurities in amounts

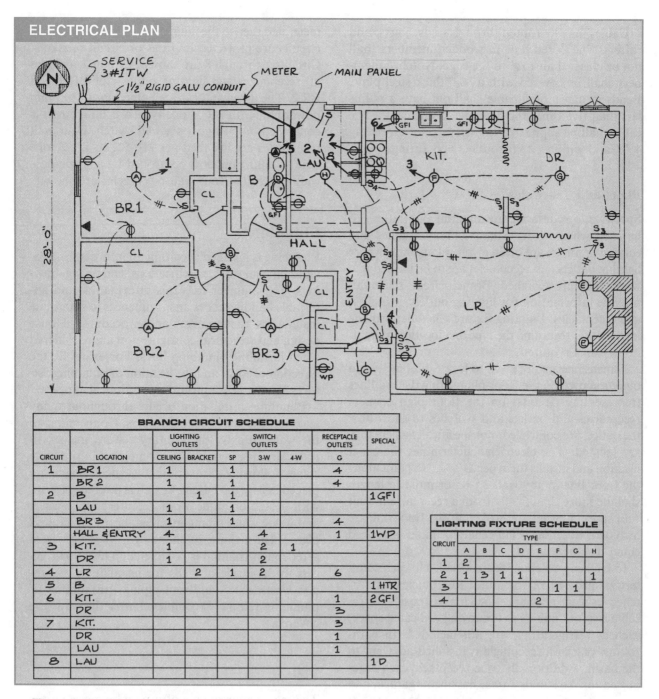

**Figure 9-11.** Floor plans show receptacles, outlets, and switches. Schedules provide additional information.

### BRANCH CIRCUIT SCHEDULE

| CIRCUIT | LOCATION | LIGHTING OUTLETS | | SWITCH OUTLETS | | | RECEPTACLE OUTLETS | SPECIAL |
|---|---|---|---|---|---|---|---|---|
| | | CEILING | BRACKET | SP | 3-W | 4-W | G | |
| 1 | BR 1 | 1 | | 1 | | | 4 | |
| | BR 2 | 1 | | 1 | | | 4 | |
| 2 | B | | 1 | 1 | | | | 1 GFI |
| | LAU | 1 | | 1 | | | | |
| | BR 3 | 1 | | 1 | | | 4 | |
| | HALL & ENTRY | 4 | | | 4 | | 1 | 1 WP |
| 3 | KIT. | 1 | | | 2 | 1 | | |
| | DR | 1 | | | 2 | | | |
| 4 | LR | | 2 | 1 | 2 | | 6 | |
| 5 | B | | | | | | | 1 HTR |
| 6 | KIT. | | | | | | 1 | 2 GFI |
| | DR | | | | | | 3 | |
| 7 | KIT. | | | | | | 3 | |
| | DR | | | | | | 1 | |
| | LAU | | | | | | 1 | |
| 8 | LAU | | | | | | | 1 D |

### LIGHTING FIXTURE SCHEDULE

| CIRCUIT | TYPE | | | | | | | |
|---|---|---|---|---|---|---|---|---|
| | A | B | C | D | E | F | G | H |
| 1 | 2 | | | | | | | |
| 2 | 1 | 3 | 1 | 1 | | | | 1 |
| 3 | | | | | | 1 | 1 | |
| 4 | | | | | 2 | | | |

that could cause harm. Supply water under pressure enters the house from a city water main or a pump. It is piped to the hot water heater and various locations of use.

**Waste Piping.** Waste piping conveys waste water and waterborne wastes from plumbing fixtures to the sanitary sewer. Sanitary sewers receive waste water that must be treated before being released into the fresh water supply. Water seals are provided on waste piping at each plumbing fixture by a trap to prevent sewer gas from entering the house. See Figure 9-13. Local building codes may require testing of the waste piping before inspection approval.

**PLUMBING SYSTEM**

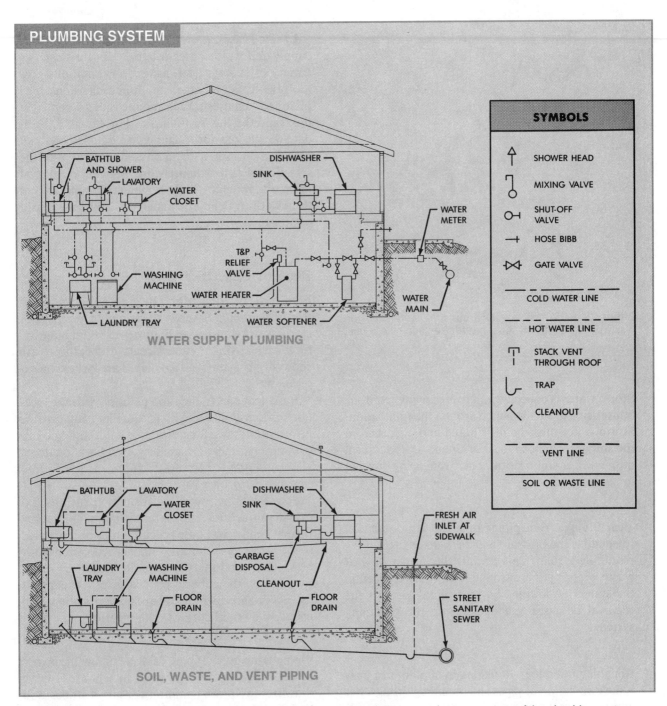

**Figure 9-12.** Supply water, waste piping, drainage piping, and vent piping are major components of the plumbing system.

In a sanitary sewer system provided by a city or other municipality, waste piping is routed directly from the house to the sewer system. In rural areas, a sanitary sewer system may not be readily available. Sanitary sewage is then routed to a septic tank and an absorption field. Solid waste collects in the septic tank and is broken down by chemical action. Waste water flows from the septic tank and is distributed into the soil of the absorption field.

**Drainage Piping.** Drainage piping carries rainwater and other precipitation (storm water) to the storm sewer or other place of disposal. This prevents collection and seepage of storm water into the house.

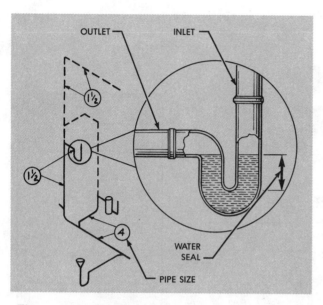

**Figure 9-13.** Traps provide a constant water seal on waste piping to prevent sewer gas from entering the house.

Storm water is collected and routed away from the house using drain tile placed around the foundation. Storm water must be distributed and absorbed in the soil according to local ordinances. Some local ordinances prohibit storm water from being directed to storm sewers.

**Vent Piping.** Vent piping provides air circulation to equalize the pressure and vacuum created in the plumbing system. This permits the proper flow of waste water by the introduction of air required in the system. Additionally, vent piping permits the removal of gases and odors from the plumbing system.

**Roughing-in Piping.** Roughing-in of plumbing lines is the installation of pipes and plumbing fittings that will be concealed after construction is completed. The exact location of roughed-in piping is critical. Specifications provide manufacturer's name and model numbers of fixtures to be installed. Dimensions for supply, waste, and vent piping required for proper installation of plumbing fixtures are provided on rough-in sheets from the manufacturer. See Figure 9-14. Specifications state that an Elkay LR 3322 kitchen sink is to be installed. The rough-in sheet shows that sink drains are 12¹/₈″ from the back edge of the sink top. The sink is 8″ deep.

In some cases, a rough-in sheet does not include supply and waste water rough-in dimensions. The variety of faucet selections or job conditions prohibit the listing of exact rough-in dimensions. The plumber then uses standard dimensions accepted in the trade for this installation. For example, standard dimensions for roughing-in a double-bowl sink specify the waste opening on center of either of the sink drains approximately 14″ above the finished floor, and roughing-in the hot and cold supply water through the wall approximately 16″ above the finished floor.

## Plans and Specifications

Plan views provide information regarding the location of plumbing fixtures, waste piping, drains, and storm water drains. See Figure 9-15. Plumbing fixtures such as sinks, water closets, and bathtubs are shown using symbols. Hose bibbs are located on exterior walls.

Basement and foundation plans provide the location of floor drains and drain tile. Plot plans provide information regarding drainage fields and hookup with supply water and sanitary sewer lines. Sectional views and interior elevations supplement information provided on plan views.

Specifications provide information such as fixture manufacturer, model numbers, and plumbing fittings required. Some specifications are included in the prints as notations. Other specifications may be sheets bound in a separate document, or forms supplied by the contractor or a federal agency. If there is disagreement between the specifications and prints, the specifications take priority.

**Piping Drawings.** Piping drawings show the layout of the entire plumbing system in a house. Piping drawings are more common for larger or custom-built houses and may not be provided in the prints. The architect, contractor, or plumber may provide piping drawings depending on the size of the job and local code requirements. Piping drawings are drawn without regard to scale or the specific location of components. Different lines are used to indicate the flow of material in the plumbing system. These lines are also used with symbols to show the installation of the required plumbing fittings. A piping drawing can be drawn in elevation view or as an isometric. See Figure 9-16.

## *Lustertone*® TWO COMPARTMENT SINK

*Lustertone*® SINK

**LR-3322**
**DOUBLE COMPARTMENT**
**18 GAUGE—TYPE 302— SELF RIM**

Highest quality, sink bowls seamlessly drawn of #18 gauge, type 302 (18-8) nickel bearing stainless steel provide convenient use from either side. Standard depth: 8-inches, with 1¾-inch radius coved corners. Self rim. Exposed surfaces are hand blended to a highlighted LK-6K-H satin finish. Underside is fully undercoated.

Model LR-3322 has a 7/16-inch drop ledge with a raised faucet deck.

Drain opening is 3½-inches.

NOTE: Where choice of three or four faucet holes are indicated, four holes furnished unless otherwise specified.

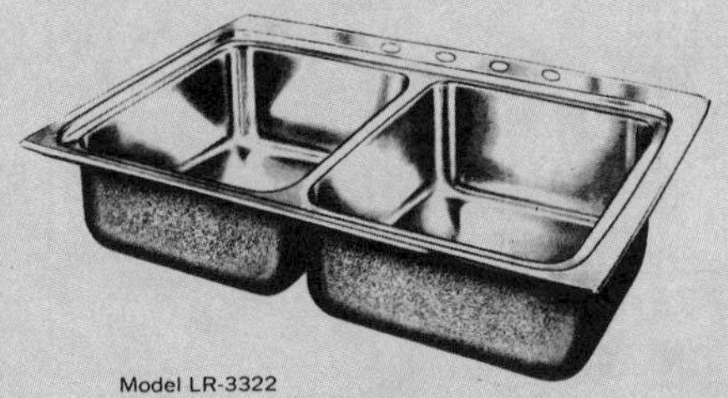

Model LR-3322

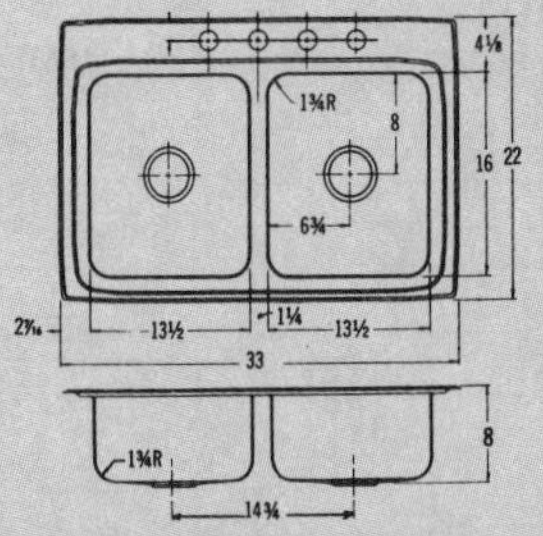


| Model Number | DIMENSIONS IN INCHES Length is left to right. Width is front to back. | | | | | Cutout In Counter Top 1½″ Radius Corners) | | No. of 1½″ Dia. Faucet Holes 4″ Centers | Ship. Wt. Lbs. |
|---|---|---|---|---|---|---|---|---|---|
| | Overall | | Inside Each Compartment | | | | | | |
| | L | W | L | W | D | L | W | | |
| **LR-3322** | 33 | 22 | 13½ | 16 | 8 | 32⅜ | 21⅜ | 3 or 4 | 21 |

**SPECIFIED FITTINGS:**

INDICATE FAUCET DRILLINGS REQUIRED

☐ 3 Holes

☐ 4 Holes

JOB NAME ______________________ DATE: __________

CUSTOMER ______________________

ARCHITECT/ENGINEER ______________________

ELKAY
stainless steel sinks

ELKAY MANUFACTURING COMPANY • 2222 CAMDEN COURT • OAK BROOK, IL. 60521

*This specification describes an Elkay product with design, quality functional benefits to the user. When making a comparison of other producers' offerings, be certain these features are not overlooked.*

*Elkay Manufacturing Company*

**Figure 9-14.** Rough-in sheets from the manufacturer provide dimensions for installation of plumbing fixtures.

## PLAN VIEWS

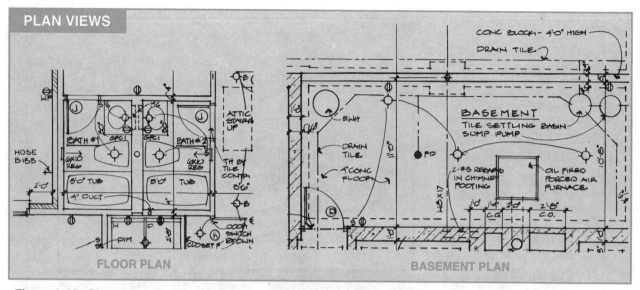

**Figure 9-15.** Plan views show the location of plumbing fixtures using symbols.

## PIPING DRAWINGS

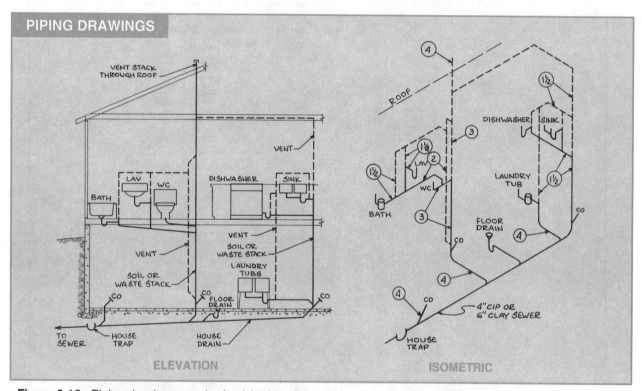

**Figure 9-16.** Piping drawings may be furnished by the architect, contractor, or plumber.

An elevation piping drawing is drawn in section with walls removed. Plumbing fixtures and piping are shown in cross section in the installed position. Isometric piping drawings are three-dimensional showing piping and plumbing fixtures drawn with vertical and 30° lines. Vertical lines represent vertical pipes. The 30° lines represent horizontal pipes. A comprehensive isometric piping drawing includes all piping and fittings for the plumbing system. The plumber uses the piping drawing to order the piping and fittings required for the job.

## HVAC

Heating, ventilating, and air conditioning (HVAC) systems supply the heat, ventilation, and cooling necessary to provide comfort to occupants of a house. The HVAC system must also control humidity, introduce fresh air, and clean and circulate air throughout the house to provide maximum comfort. A perceived change in any of these requirements will reduce the level of comfort.

HVAC installation occurs after rough framing and before final finish of walls. The installation of the HVAC system involves several trade areas. Ductwork is installed by sheet metal workers. Piping for hot water heating systems and gas lines are installed by plumbers. Electricians install wiring for the HVAC system. Cement masons perform the cement work for the installation of HVAC equipment. Carpenters work closely with the other trades to provide proper access for piping, electrical connections, and ductwork. Each trade must review the prints for information pertinent to the specific trade. The three types of residential heating systems commonly used are forced warm air, hot water (hydronic), and radiant electric. Alternate heating systems are used in regions where feasible.

## Forced Warm Air Heating

*Forced warm air heating* uses a blower to draw air (return air) from rooms through return air grills and ductwork. The return air passes through a furnace where it is heated. The heated air (supply air) passes through the plenum and is distributed to the desired rooms through supply ductwork and registers. Supply air ducts are commonly placed between floor joists or under slab-on-grade floors. Supply air ducts branch off from a trunk supply line centrally located in the house. Supply air registers are generally located close to the exterior walls of the house with return air grills and ductwork located near the center of the house. This arrangement allows the required circulation of return air from rooms to the furnace and supply air back to the rooms. See Figure 9-17.

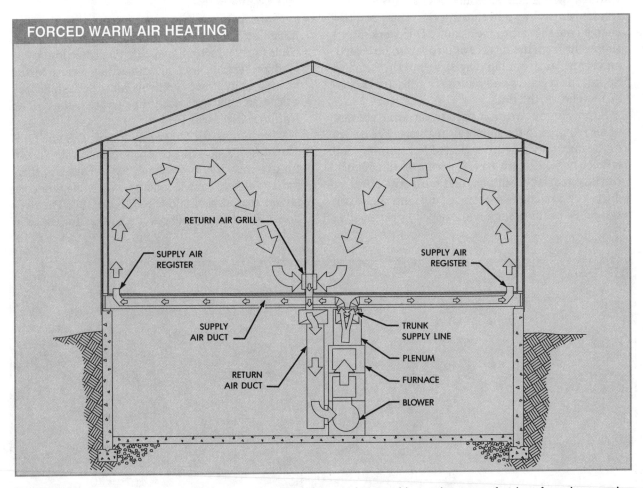

**FORCED WARM AIR HEATING**

**Figure 9-17.** Supply air registers and return air grills are located to provide maximum comfort in a forced warm air heating system.

The major components of a forced warm air heating system are the furnace, blower, ductwork, and unit controls. *Furnaces* heat air to be circulated in the forced warm air heating system. They differ in design and operation based on how heat is generated. Fuels such as fuel oil or natural gas, or electricity are commonly used to generate heat in a furnace. Furnaces using fuel have a burner, combustion chamber, heat exchanger, and unit controls. The combustion process takes place in the combustion chamber with air heated as it passes through the heat exchanger. Furnaces using electricity have resistance heating elements in place of the burner, combustion chamber, and heat exchanger.

*Unit controls* operate the individual parts of the furnace and control the amount of heat produced. *Blowers* move air from rooms, through the furnace where the air is heated, and back into the rooms. The most common blower type is the centrifugal blower. It has a blower wheel that rotates rapidly within a sheet metal enclosure. As the blower wheel rotates, air is drawn in through the center and discharged out the side to the plenum. Ductwork directs supply air from the furnace into rooms to be heated, and return air from building spaces to the furnace. Supply air registers and return air grills are located at the end of the ducts.

Supply air ductwork is sized to carry the amount of air required to the supply air registers. The design of the ductwork system is selected for economy and efficiency. See Figure 9-18. A perimeter radial ductwork system is the simplest and most cost-effective ductwork system. An extended plenum ductwork system is useful in long houses such as ranch-style houses. A perimeter loop ductwork system is commonly used in cold climates. Ductwork is made from galvanized sheet metal, fiberglass, and reinforced aluminum-faced materials. All ductwork must comply with applicable codes.

Supply air registers are located on the supply ducts to supply heated air to the desired room. Supply air registers are sized to supply and direct the proper amount of heated air to the desired room for maximum efficiency. See Figure 9-19. Return air grills in forced warm air heating systems are located and sized to return as much air from building spaces as supplied. It is common to have fewer return air grills than supply air registers. However, return air grills are usually larger and more centrally located than supply registers.

*Unit controls* are controls installed on a furnace by the manufacturer or installer to maintain safe and efficient operation of the furnace. Unit controls are classified as power controls, operating controls, and safety controls.

*Power controls* are controls located in the electrical lines leading to equipment. Power controls include the conductors, overcurrent protection, overload protection, and disconnecting means to the heating unit. Power controls must be installed by a licensed electrician according to provisions of the National Electrical Code®.

*Operating controls* cycle equipment ON and OFF as required. Operating controls include a step-down transformer in a low-voltage control system, thermostat, blower control, and operating relays or contactors that turn components ON or OFF. The step-down transformer reduces the voltage for usage in

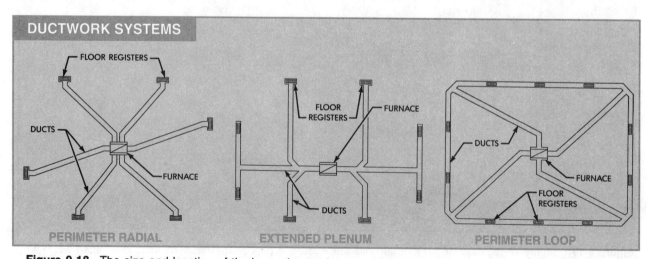

**Figure 9-18.** The size and location of the house have a bearing on the type of ductwork system required.

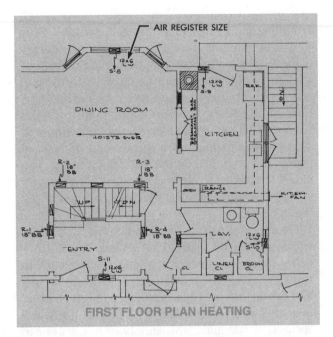

**Figure 9-19.** The size of supply air registers is specified on the floor plan.

the control system. The thermostat is a temperature-sensing electrical switch that turns the furnace ON and OFF. The blower control operates the blower used to circulate air through the furnace and the desired rooms.

*Safety controls* prevent injury to personnel or damage to equipment in the event of an equipment malfunction. Safety controls include such controls as limit and combustion safety controls. Motor overloads may also be required in some furnaces.

## Hot Water Heating

*Hot water heating* heats water in a boiler at a central location. The heated water is distributed to the desired rooms by a circulating pump through supply water piping. At the rooms to be heated, the hot water passes through terminal devices. Terminal devices extract heat from the hot water to heat air in the desired rooms. The water then returns to the boiler through return water piping. Hot water heating systems include a boiler, circulation pumps, piping system, terminal devices, and controls. A compression tank relieves pressure caused by water expansion when heated. See Figure 9-20. Most hot water systems are central systems with one boiler used for heating water to be distributed to the required rooms through a piping system.

Boilers used to heat water are classified according to how heat is generated. Fuels commonly used include natural gas and fuel oil. Electric heating elements are also used if other fuels are unavailable or less cost efficient. Water heated by the sun (solar) or hot water from the ground (geothermal) can also be used, if feasible.

Circulating pumps move water through the hot water heating system. The circulating pump is commonly located in the return water piping close to the boiler. The size of circulating pump required is determined by the volume of water in the system. Piping is used to distribute hot water from the boiler to the terminal devices and return water back to the boiler to be reheated. Terminal devices commonly used include radiators, convectors, blower coils, and radiant panels. A fan or blower may be used to circulate the heated air more efficiently.

Hot water heating system piping may also be installed in a slab-on-grade floor. Pipes are set in place before the concrete is poured. Heat from the hot water is radiated through the concrete. Unit controls control the boiler, circulating pump, and terminal devices and are similar to unit controls for a forced warm air heating system.

Hot water heating systems in residential installations most commonly use a one-pipe system. Hot

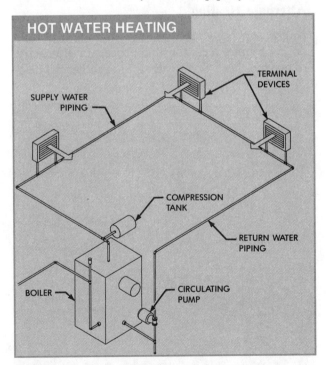

**Figure 9-20.** The circulating pump is located in the return water piping and provides the pressure required to pump heated water through the system.

water is distributed to the desired rooms using branches and risers. The amount of hot water flowing through each terminal device is controlled by a valve setting or a thermostat. Two-pipe systems use a separate pipe for supply and return water. This provides a more consistent water temperature to all terminal devices as cooled water is returned in a separate pipe to the boiler. The two-pipe system requires additional piping and often is too costly for most residential construction.

## Radiant Electric Heating

*Radiant electric heating* generates heat as electricity meets resistance when passed through a conductor. Radiant electric heating is used in panel systems and baseboard units. Panel systems have thin wires embedded in prefabricated floor, wall, or ceiling finish materials. See Figure 9-21. The amount of electricity flowing through the wires in the panels is controlled by a thermostat. Baseboard units are installed after the walls are finished and are mounted to the wall or floor. Air is heated as it contacts the heated electric coils. Some baseboard units contain water to extend the time that heat is retained and transferred to the air.

## Alternate Heating

*Alternate heating* uses natural sources of heat that do not require a combustion process or electricity to produce heat. Solar energy and geothermal heat

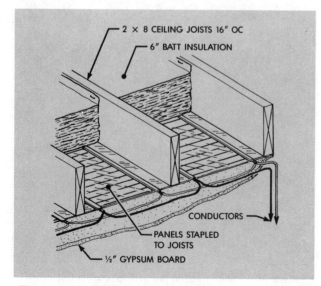

**Figure 9-21.** Radiant electric heating panels have thin wires embedded within each panel that radiate heat.

from thermal processes in the earth are two of the most common alternate heat sources. *Solar energy* is the energy available from the sun in the form of sunlight. Solar energy can be changed to thermal energy used to heat air and water for heating purposes. Solar energy systems collect and store solar energy. Well-designed solar energy collection and storage systems maximize the amount of thermal energy received. See Figure 9-22. Solar energy can also be converted to electrical energy. The electricity is then used for operating heating or air conditioning systems.

*Geothermal heat* is heat that comes from deep within the earth's crust. The center of the earth is believed to consist of hot molten rock (*magma*). In some locations, groundwater that sinks into the soil and subsurface rock reaches the magma. The water is heated and returns to the surface as hot water springs. If adequate, this supply of hot water can be used for heating purposes. In addition to natural sources, deep wells and pumps are also used to pump hot water to the surface.

## Air Conditioning

*Air conditioning* cools, filters, and dehumidifies air. The temperature desired is controlled by a thermostat. Cooled air is circulated through the desired rooms where it absorbs heat and is returned to the air conditioning unit. The two types of air conditioning systems commonly used in residential construction are the split system and the package system. See Figure 9-23.

In the *split system,* the two main components (condensing unit and evaporator) are split into two locations. The condensing unit is located on a concrete slab outside of the house. The evaporator (cooling coil) is located on the discharge side of the furnace blower. Heat is absorbed as it passes through the fins of the cooling coil. A condensate drain line directs water that has condensed from excess humidity in the air to the drain. The ductwork used in the forced warm air heating system is used to transport the cooled air. When designing the HVAC system using a split system, the additional resistance to flow caused by the cooling coil must be factored.

*Package systems* have the major components (condensing unit and evaporator coil) built into the same unit. Supply air and return air are ducted to the unit. The unit can be mounted in the attic, in

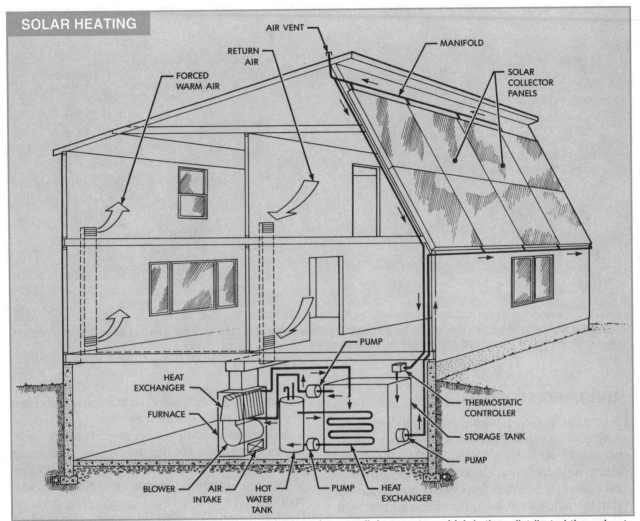

**SOLAR HEATING**

AIR VENT

RETURN AIR

FORCED WARM AIR

MANIFOLD

SOLAR COLLECTOR PANELS

PUMP

HEAT EXCHANGER

FURNACE

THERMOSTATIC CONTROLLER

STORAGE TANK

PUMP

BLOWER

AIR INTAKE

HOT WATER TANK

PUMP

HEAT EXCHANGER

**Figure 9-22.** Solar energy heating systems transfer heat from sunlight to water, which is then distributed throughout the house for heating purposes.

the garage, or through foundation walls. Some package air conditioning systems have the heating system components built in.

## Plans and Specifications

The plans used by HVAC installers include the plot plan, foundation plan, and floor plan. The plot plan shows the location of the air conditioner and the natural gas supply. The foundation plan shows information regarding ventilation fan locations, furnace, boiler, air conditioning equipment locations, and ductwork. The floor plan gives the size and location of supply ducts and return ducts. If required, furnace, boiler, and air conditioning information is also provided on the floor plan. The direction of floor joists and studs in partitions and

walls must be determined before piping or ductwork can be run. Information regarding the HVAC system is provided using dimensions and symbols. Refer to the Appendix.

Additional information may be provided on mechanical schedules, ductwork layouts provided by the architect or the HVAC contractor, and the specifications. An engineer may be contracted by the architect or contractor to determine the specific equipment, ductwork, and/or piping requirements of the system.

Specifications for the HVAC system include information that is not included in other parts of the prints. HVAC equipment is described in the specifications by the manufacturer's name, model number, and type of system. The output capacity and materials used are also included.

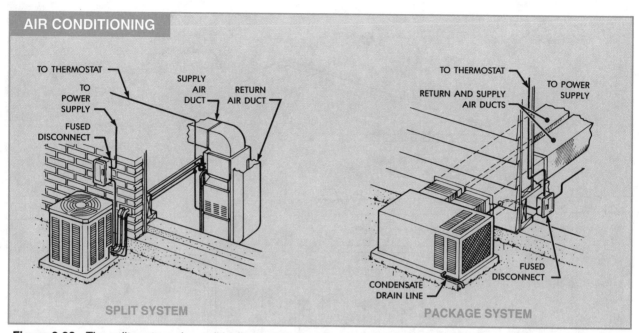

AIR CONDITIONING

TO THERMOSTAT
TO POWER SUPPLY
FUSED DISCONNECT
SUPPLY AIR DUCT
RETURN AIR DUCT

SPLIT SYSTEM

TO THERMOSTAT
RETURN AND SUPPLY AIR DUCTS
TO POWER SUPPLY
FUSED DISCONNECT
CONDENSATE DRAIN LINE

PACKAGE SYSTEM

**Figure 9-23.** The split system air conditioning system has the condensing unit in a different location from the cooling coil. Components of a package system are in the same unit.

## SHEET METAL WORK

Sheet metal workers fabricate and install sheet metal products and other sheet materials to be used on the outside and inside of the house. On the outside, sheet metal products are used for providing protection against the weather. Storm water is routed away from the house by gutters, flashing, and trim. On the inside, the majority of sheet metal work required is done on the HVAC ductwork system.

Many sheet metal components are prefabricated. Some sheet metal applications require the product to be fabricated on site. This requires a knowledge of pattern development, cutting, and bending the materials into the required shape.

Information regarding sheet metal work required on the job is included in foundation plans, exterior elevation views, floor plans, and detail views. Additional information may also be provided in the specifications, or by the architect or contractor. On the outside of the house, termite shields shown on the wall section are used in regions with large insect populations. Termite shields are becoming less of a necessity in most regions as treated wood serves as an adequate barrier against termites.

Sheet metal work is required when installing roofing to ensure protection and proper storm water flow. Applications on roof installations include flashing and gravel stops. Flashing required around

a chimney is determined by the location of the chimney and the pitch of the roof.

A sheet metal saddle is fabricated to prevent water from seeping under roofing materials. Step flashing is overlapped on the sides of the chimney. Flashing is required around structures extending from the roofline, such as a dormer. Overlapping strips of flashing material are used. Flashing is also used on built-up roofing between vertical and horizontal surfaces. Gutters or special ornamental vents may also be installed by sheet metal workers on the job. See Figure 9-24.

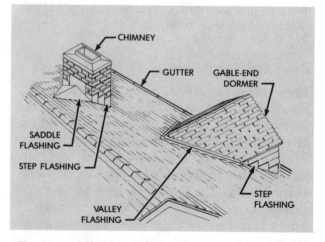

CHIMNEY
GUTTER
GABLE-END DORMER
SADDLE FLASHING
STEP FLASHING
STEP FLASHING
VALLEY FLASHING

**Figure 9-24.** Sheet metal work provides protection against the weather.

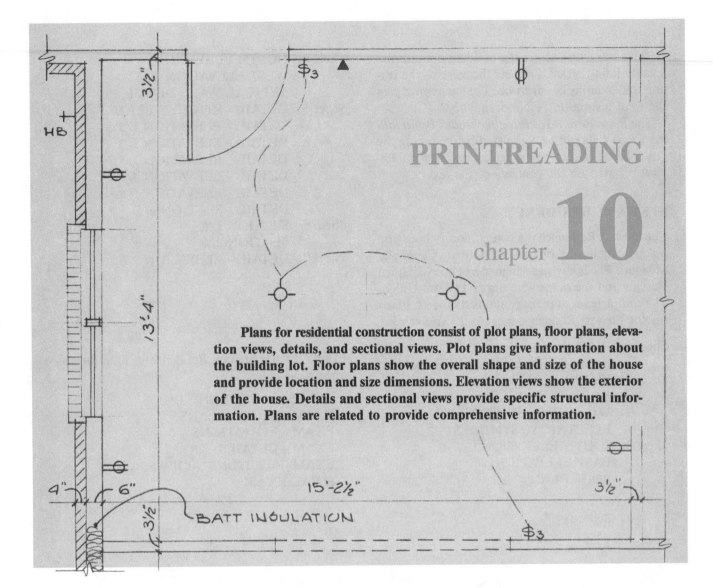

# PRINTREADING

chapter **10**

**Plans for residential construction consist of plot plans, floor plans, elevation views, details, and sectional views. Plot plans give information about the building lot. Floor plans show the overall shape and size of the house and provide location and size dimensions. Elevation views show the exterior of the house. Details and sectional views provide specific structural information. Plans are related to provide comprehensive information.**

## PRINTREADING

Plans for residential construction are drawn following accepted drafting practices. They may be drawn by the conventional method or CAD method. Prints are produced from the working drawings. The prints and specifications are used by contractors and subcontractors to determine bids. Tradesworkers refer to the prints throughout construction to determine materials, location dimensions, and size dimensions.

Symbols and abbreviations are used to standardize plans and conserve space. Floor plans are generally the first sheet(s) to be drawn. The most commonly used scale for drawing floor plans is $\frac{1}{4}'' = 1'-0''$. Elevation views show the exterior of the house. The most commonly used scale for drawing elevation views is also $\frac{1}{4}'' = 1'-0''$. Cutting planes on plan views refer to sectional views and details. Sectional

views provide additional information. Details are drawn to larger scales to clearly show the type and size of construction material required to complete the house. Window and door schedules provide specific type and size of windows and doors needed. Windows and doors are referenced to the schedules. Plot plans show the size and shape of the lot and the location of the house on the lot. Utilities and streets are also shown. Plot plans are drawn to smaller scales. For example, a common scale for residential plot plans is $1'' = 20'-0''$.

The work of construction workers in the various trades is coordinated to facilitate construction. Many trades complete their rough-in and then return to complete the job. For example, electricians rough-in all circuits after stud walls are erected and install receptacles, switches, and so forth after finished walls are in place.

The contractor secures the building permit, and local building officials inspect the work of the various trades during construction. The house must pass the final inspection before occupancy.

The Final Review for *Building Trades Printreading - Part 1* should be completed before taking the exams based on the Stewart Residence. The Final Review is based on the contents of the text.

## STEWART RESIDENCE

The Stewart Residence is a contemporary house on a sloping lot located on Leawood Drive in Columbia, Missouri. The house was designed by Hulen & Hulen Designs and the plans were drawn by Pam Hulen. The complete set of plans contains six sheets. Plans for the Stewart Residence include the following:

Sheet 1:  PLOT PLAN
          BASEMENT LEVEL FLOOR PLAN
Sheet 2:  FLOOR PLAN
          DOOR SCHEDULE
          WINDOW SCHEDULE
          ELECTRICAL SYMBOLS
Sheet 3:  SOUTH ELEVATION
          EAST ELEVATION
          ROOF PLAN

Sheet 4:  NORTH ELEVATION
          WEST ELEVATION
          TYPICAL WALL DETAIL
Sheet 5:  DETAIL - KITCHEN PLAN
          DETAIL - ELEVATION 1/5
          DETAIL - ELEVATION 2/5
          DETAIL - ELEVATION 3/5
          DETAIL - ELEVATION 4/5
          DETAIL - ELEVATION 5/5
          DETAIL - SECTION 6/5
Sheet 6:  SECTION 1/6
          SECTION 2/6
          DETAIL - ELEVATION 3/6

Exams for the Stewart Residence include the following:

EXAM—PLOT PLAN
EXAM—FLOOR PLANS
EXAM—ELEVATIONS
EXAM—DETAILS
EXAM—SECTIONAL VIEWS
FINAL EXAM

**Final Review**
Questions

Name _____ Date _____

_____ 1. A township contains _____.
   A. 36 square miles
   B. 36 quarter sections
   C. both A and B
   D. neither A nor B

_____ **2.** The denominator of a fraction is the number _____ the fraction bar.
   A. to the left of
   B. to the right of
   C. above
   D. none of the above

T   F   **3.** The size of floor joists is generally given on elevation views.

T   F   **4.** Rebars increase the strength of concrete.

T   F   **5.** Elevations are pictorial drawings showing interior views of a building.

_____ **6.** A dimension showing the distance from a building corner to the center of the rough opening for a window is a _____ dimension.
   A. size
   B. location
   C. both A and B
   D. neither A nor B

_____ **7.** The abbreviation 7 R DN refers to the number of _____ descending to a lower floor level.

T   F   **8.** A true blueprint has white lines on a blue background.

_____ **9.** Studs in _____ framing run full length from the sill to the double top plate.

_____ **10.** The North Elevation is the _____.
   A. elevation facing North
   B. direction a person faces to see the North side of the house
   C. both A and B
   D. neither A nor B

_____ **11.** Solid bridging is generally _____ to facilitate nailing.

T   F   **12.** Cutting planes for floor plans are taken 6'-0" above the finished floor.

T   F   **13.** True North is designated on the plot plan to show house orientation on the lot.

_____ **14.** The size of a window may be given _____.
   A. on the elevation views
   B. in the window schedule
   C. both A and B
   D. neither A nor B

T     F     **15.** The slope of the roof is shown on exterior elevations.

_____     **16.** Detail views are _____ views drawn to a larger scale to show additional information.
A. sectional
B. elevation
C. plan
D. all of the above

_____     **17.** The _____ size of a piece of wood is its size before planing.

T     F     **18.** Quadrilaterals are three-sided plane figures.

_____     **19.** The most common diazo process utilizes _____ to produce prints.
A. chlorine
B. ammonia
C. both A and B
D. neither A nor B

T     F     **20.** The radius of a circle is twice the length of the diameter.

_____     **21.** The _____ is the lowest wooden member in platform framing.

_____     **22.** Regarding CAD, _____.
A. monochrome monitors are preferred
B. layering facilitates the drawing of plans for specific trade areas
C. a mouse is an output system
D. none of the above

T     F     **23.** The broad aspects of shape, size, and relationship of rooms are shown on floor plans.

_____     **24.** Detail 4/7 is found on Sheet _____.

T     F     **25.** A riser is the vertical portion of a stairstep.

T     F     **26.** A two-story house with basement requires two floor plans.

_____     **27.** Regarding angles, a _____.
A. right angle contains 90°
B. straight line contains 180°
C. both A and B
D. neither A nor B

_____     **28.** The point of beginning is the location point for _____ dimensions.
A. horizontal
B. vertical
C. both A and B
D. neither A nor B

T     F     **29.** An isosceles triangle contains one 90° angle.

T     F     **30.** Exterior doors must be at least 1¾" thick.

_____     **31.** Kitchen base cabinets are commonly _____" high and _____" deep.
A. 24, 30
B. 30, 36
C. 36, 24
D. 36, 30

_____     **32.** In an oblique _____ drawing, receding lines are drawn at one-half scale.

_____ 33. The location of streets, easements, and utilities is shown on the _____.
    A. sectional views
    B. details
    C. elevation views
    D. none of the above

T   F   34. GFI receptacles must be placed adjacent to kitchen sinks.

T   F   35. A gable roof has a double slope in two directions.

_____ 36. The _____ scale is used to draw plot plans.
    A. architect's
    B. civil engineer's
    C. mechanical engineer's
    D. none of the above

T   F   37. Air conditioning units remove moisture from air in the house.

T   F   38. In relation to roofs, run is the overall distance between building corners.

_____ 39. Natural grade is _____.
    A. the slope of the lot before rough grading
    B. the slope of the lot after rough grading
    C. shown on plot plans with solid lines
    D. none of the above

_____ 40. A(n) _____ section is made by passing a cutting plane through the short dimension of a house.

_____ 41. Dimensions may be terminated by _____.
    A. arrowheads
    B. slashes or dots
    C. both A and B
    D. neither A nor B

T   F   42. The preferred method of dimensioning framed walls is to dimension to the outside face of stud corner posts.

T   F   43. Slope is the relationship of roof rise to run.

T   F   44. All triangles contain 180°.

_____ 45. The architect's scale _____.
    A. is graduated in decimal units
    B. contains a ruler and nine scales
    C. may be triangular in shape
    D. none of the above

T   F   46. A lateral electrical service is run underground.

T   F   47. The minimum size branch circuit for a dwelling is 15 A.

_____ 48. Size dimensions for rooms are most clearly shown on _____.
    A. plot plans
    B. floor plans
    C. sectional views
    D. details

_____ 49. _____ framing is the most common type of framing for residential construction.

T    F    **50.** A light is a pane of glass.

T    F    **51.** Floor joists may be lapped together.

T    F    **52.** Concrete block partitions are generally dimensioned to the face of the finish material.

_____    **53.** A line 2¼″ long on a plan drawn at the scale of ¾″ = 1′-0″ represents a dimension of _____ in the finished house.

T    F    **54.** The apex of a triangle drawn on a window in an elevation view points to the hinged side.

T    F    **55.** Termite shields or treated wood may serve as a barrier against termites.

T    F    **56.** A 15 A, 120 V circuit contains 8 VA.

T    F    **57.** Supply air registers of a forced warm air heating system are generally located near the center of a house.

_____    **58.** The scale most commonly used on floor plans is _____.
     A. ⅛″ = 1′-0″
     B. ¼″ = 1′-0″
     C. ⅜″ = 1′-0″
     D. ½″ = 1′-0″

_____    **59.** Compass directions are commonly used to _____.
     A. name exterior elevation views
     B. locate dimensions for rough openings of doors and windows
     C. denote roof slope
     D. determine swale on building lots

_____    **60.** Horizontal lines of isometric drawings are drawn _____° above the horizon.
     A. 15
     B. 30
     C. 45
     D. 90

## Identification 10-1

_____  1. Concrete (Elevation View)

_____  2. Glass (Elevation View)

_____  3. Rough framing member

_____  4. Sliding doors (Plan View)

_____  5. Single-bowl vanity (Plan View)

_____  6. Common brick (Plan View)

_____  7. Ceiling light

_____  8. Hose bibb

_____  9. Brick veneer (Plan View)

_____ 10. Switch

## Identification 10-2

_____  1. Isosceles triangle

_____  2. Hexagon

_____  3. Acute angle

_____  4. Square

_____  5. Equilateral triangle

_____  6. Eccentric circles

_____  7. Concentric circles

_____  8. Rectangle

_____  9. Rhombus

_____ 10. Octagon

(D) (two equal sides, two equal angles)

(E) (three equal sides, three equal angles)

(F) (eight sides)

(G) (six sides)

(H) (four equal sides, opposite angles equal)

(I) (opposite sides equal, four 90° angles)

(J) (four equal sides, four 90° angles)

(C) LESS THAN 90°

## Identification 10-3

_____  **1.** Baseboard

_____  **2.** Vertical rebar

_____  **3.** Horizontal rebar

_____  **4.** Sill

_____  **5.** Floor truss

_____  **6.** Brick

_____  **7.** Foundation footing

_____  **8.** Brick tie

_____  **9.** Foundation wall

_____  **10.** Earth

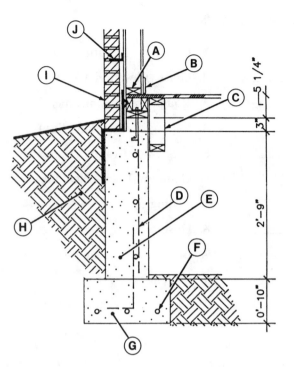

## Completion

_____  **1.** ¼″ = 1′-0″

_____  **2.** ¼″ = 1′-0″

_____  **3.** ¼″ = 1′-0″

_____  **4.** ¼″ = 1′-0″

_____  **5.** ⅜″ = 1′-0″

_____  **6.** ⅜″ = 1′-0″

_____  **7.** ½″ = 1′-0″

_____  **8.** 1″ = 1′-0″

_____  **9.** 1½″ = 1′-0″

_____  **10.** 3″ = 1′-0″

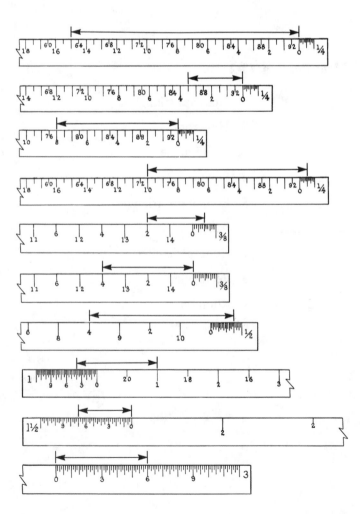

# Exam—Plot Plan
*Review Questions*

Name _____ Date _____

_____ **1.** The Plot Plan is drawn to the scale of _____.

_____ **2.** The lot for the Stewart Residence generally slopes down from _____ to _____.
    A. Southeast, Northwest
    B. Northeast, Southwest
    C. Northwest, Southeast
    D. the lot has no slope

T    F **3.** PLH drew the Plot Plan for the Stewart Residence on 3-17-89.

_____ **4.** The lot for the Stewart Residence contains _____ sq ft.
    A. 514
    B. 11,025
    C. 15,960
    D. 23,104

_____ **5.** The front of the Stewart Residence faces _____.

_____ **6.** The finished floor elevation of the main level is _____ above the finished floor elevation of the garage.

T    F **7.** A 24″ existing pine tree in the front yard has a 20′-0″ diameter drip.

_____ **8.** The Southeast corner of the garage is _____ from the point of beginning and _____ from the East property line.
    A. 15′-0″, 40′-0″
    B. 40′-0″, 15′-0″
    C. 40′-0″, 96′-0″
    D. 105′-0″, 152′-0″

_____ **9.** The general shape of the lot for the Stewart Residence is a _____.
    A. rhombus
    B. rhomboid
    C. trapezoid
    D. trapezium

T    F **10.** The den is located on the South side of the Stewart Residence.

T    F **11.** Leawood Drive is parallel to the East property line of the Stewart Residence.

_____ **12.** The Northeast corner of the lot is approximately _____ above the point of beginning.
    A. 6′-0″
    B. 8′-0″
    C. 10′-0″
    D. 12′-0″

_____ 13. The lot for the Stewart Residence is _____ wide and _____ deep measured along the property lines.
   A. 84'-0", 96'-0"
   B. 96'-0", 105'-0"
   C. 105'-0", 152'-0"
   D. none of the above

_____ 14. The finished floor elevation of the den is _____.

_____ 15. A _____ on the finished grade diverts surface water away from the Northwest corner of the Stewart Residence.

_____ 16. A deck is located on the _____ side of the Stewart Residence.

_____ 17. The driveway slopes approximately _____ from the garage entrance to Leawood Drive.
   A. 2'-0"
   B. 4'-0"
   C. 6'-0"
   D. 8'-0"

_____ 18. An easement for utilities is located along the _____ property line.

T   F    19. Three existing trees are shown on the lot for the Stewart Residence.

_____ 20. The elevation for the point of beginning is _____.

_____ 21. The lot for the Stewart Residence is steepest near the front _____.

_____ 22. The finished floor elevation of the garage is _____.

_____ 23. The West property line is _____ long.

_____ 24. South and East property lines meet at an angle of _____.

_____ 25. The Stewart Residence is set back _____ from the front property line.

# Exam—Floor Plans

*Review Questions*

Name _____ Date _____

_____ 1. The overall dimensions of the Stewart Residence, not including the deck, are 42'-5" × _____.

_____ 2. Floor plans for the Stewart Residence are drawn to the scale of _____.

_____ 3. The main electrical panel is located in the _____.

_____ 4. The future bath is stubbed in for a _____.
   A. lavatory, water closet, and shower
   B. lavatory, water closet, and tub
   C. double-bowl vanity, water closet, and tub
   D. no future bath is shown

_____ 5. Notations on the Basement Level Floor Plan indicate that floor joists for the main level are _____.

T   F   6. The incandescent fixture in the dining room is controlled by three-way switches.

_____ 7. Regarding the bedrooms, _____.
   A. Bedroom 3 has a walk-in closet
   B. Bedroom 2 is larger than Bedroom 3
   C. cable TV outlets are located in each bedroom
   D. the master bedroom has four recessed can fixtures

_____ 8. Regarding the den, _____.
   A. bypass doors separate the den from the entry
   B. bay windows are located in the West wall
   C. overall room dimensions are 13'-0" × 15'-8½"
   D. six duplex outlets are located 12" above the finished floor

_____ 9. Regarding the kitchen, _____.
   A. a 3 ft sq skylight is located in the panned ceiling
   B. the refrigerator is located on the North wall
   C. casement windows provide a view of the front yard
   D. the cooktop is mounted in the island cabinet

_____ 10. Regarding the living room, _____.
   A. built-in cabinets and shelves flank the fireplace
   B. all duplex outlets are located 12" above the finished floor
   C. ceiling lighting is controlled by three-way switches
   D. casement windows with a circle top are located in the East wall

_____ 11. Regarding the dining room, _____.
   A. sliding glass doors lead to the deck
   B. a ceiling fan with light fixture is controlled by a wall switch
   C. both A and B
   D. neither A nor B

_____ 12. Regarding the entry, _____.
    A. a coat closet is to the right of the front door
    B. a stairway with 7 R UP leads to the main level
    C. four skylights provide natural light
    D. a 12″ sidelight is located on the hinged side of the door

_____ 13. Regarding the bathrooms, _____.
    A. both baths on the main level are the same size
    B. the master bath contains a linen closet
    C. a fan light fixture is wired in the future bath
    D. none of the above

_____ 14. Regarding the future family room, _____.
    A. a full glass door leads to the deck
    B. a telephone outlet is adjacent to the fireplace
    C. the overall size is 14′-9½″ × 17′-4½″
    D. no windows are shown in the West wall

_____ 15. Regarding the future workshop, _____.
    A. casement windows provide natural light
    B. floor joists above run East and West
    C. all duplex outlets are 42″ above the finished floor
    D. none of the above

_____ 16. Regarding the garage, _____.
    A. a 7′-0″ × 16′-0″ overhead door is shown
    B. the lower level of the house is three steps up
    C. an overhead steel beam is supported by 3″ diameter steel columns
    D. overall dimensions are 13′-1¼″ × 23′-2½″

T    F    17. A low storage area is located beneath the den.

T    F    18. Two cutting planes on the Floor Plan refer to sectional views shown on Sheet 6.

_____ 19. A total of _____ telephone outlets are installed in the Stewart Residence.

_____ 20. The hallway on the main level is _____ wide.

_____ 21. _____ lighting fixtures in the kitchen are controlled by three-way switches.

T    F    22. The walk-in closet in the master bedroom has two shelves with rods.

T    F    23. The lighting fixture in the storage closet of the future family room is controlled by a pull chain.

T    F    24. The concrete slab beneath the stairway is thickened to support the additional weight.

_____ 25. The finished floor level of the low storage area is _____″ above the finished floor level of the garage.

_____ 26. A(n) _____ fixture near the fireplace in the future family room is controlled by a wall switch.

_____ 27. The attic is entered through a scuttle measuring _____.

_____ 28. The deck measures 10′-0″ × _____.

_____ 29. Closets in Bedrooms 2 and 3 are _____ deep.

| | | |
|---|---|---|
| _____ | **30.** | The linen closet is _____ wide. |
| _____ | **31.** | A panned ceiling is shown in the _____. |
| _____ | **32.** | The ceiling in the _____ room is vaulted. |
| _____ | **33.** | The concrete foundation wall beneath the den is _____ " thick. |
| T   F | **34.** | The water heater is located in the laundry room. |
| _____ | **35.** | The North wall of the kitchen projects _____ beyond the rest of the North wall of the Stewart Residence. |
| T   F | **36.** | The hall bath contains two GFI outlets. |
| T   F | **37.** | Bay windows are located in the living room. |
| T   F | **38.** | The front stairway is 4'-6" wide. |
| T   F | **39.** | The linen closet in the master bath has no lighting fixture. |
| T   F | **40.** | Two garage door openers are shown on the garage ceiling. |
| _____ | **41.** | A total of _____ duplex outlets are located 12" above the finished floor in the walls of the future family room. |
| _____ | **42.** | H windows in the future family room are centered _____ from the Northwest corner of the house. |
| _____ | **43.** | E windows in the workshop are centered _____ from the Northeast corner of the house. |
| T   F | **44.** | A floor drain is shown in the center of the garage floor. |
| T   F | **45.** | Two H windows provide natural light in the laundry room. |
| T   F | **46.** | The hallway on the main floor is 21'-1¼" long. |
| T   F | **47.** | Ceiling fixtures on the kitchen ceiling are fluorescent. |
| _____ | **48.** | The service-entrance conductors enter the house on the _____ side. |

_____ **49.** Bedroom 3 contains approximately _____ sq ft.
    A. 93
    B. 103
    C. 133
    D. 153

_____ **50.** The master bedroom contains approximately _____ sq ft.
    A. 117
    B. 137
    C. 157
    D. 177

| | | |
|---|---|---|
| _____ | **51.** | The storage closet beneath the stairway measures 3'-5¾" × _____. |
| _____ | **52.** | D windows in the den are centered _____ from the framed Southwest corner. |
| T   F | **53.** | A cased opening with arch leads from the kitchen to the dining room. |
| _____ | **54.** | The main entry is _____ wide. |
| _____ | **55.** | The bay for A windows in the den is _____ wide. |

_____ **56.** Four recessed can lighting fixtures are shown in the _____.
  A. den
  B. master bedroom
  C. both A and B
  D. neither A nor B

T  F  **57.** Wrought iron handrails are shown for the entry steps.

T  F  **58.** Refrigerator space in the kitchen is provided on the East wall.

T  F  **59.** All bedroom closets have bypass doors.

T  F  **60.** The wall receptacle on the North wall of the living room is split-wired.

# Exam—Elevations

Review Questions

Name _____ Date _____

_____ 1. All elevations are drawn to the scale of _____.

_____ 2. The _____ Elevation shows the front of the Stewart Residence.

_____ 3. Regarding the South Elevation, _____.
   A. the fireplace chimney is centered on the bay windows
   B. sidelights flank the entry door
   C. all exterior walls are brick veneer
   D. one skylight is centered on the entry door

_____ 4. Regarding the East Elevation, _____.
   A. the roof slopes 4 in 12
   B. a CCA treated handrail leads to the entry door
   C. exterior walls are brick veneer and 4″ lap siding
   D. a 3′-0″ diameter glass block window and one-half glass door provide natural light in the garage

_____ 5. Regarding the North Elevation, _____.
   A. the deck extends completely across the back side of the Stewart Residence
   B. 4 × 4 treated posts set in 12″ diameter concrete piers support the deck
   C. exterior walls are brick veneer and 8″ lap siding
   D. the roof is finished with cedar shakes

_____ 6. Regarding the West Elevation, the _____.
   A. den floor and basement floor are at the same level
   B. chimney is brick veneer with a 4″ limestone cap
   C. roof slope is 5 in 12
   D. none of the above

T   F   7. The deck floor level is level with the main floor level.

T   F   8. Overhead garage doors are vinyl-clad metal.

_____ 9. A total of _____ C windows are shown in the East Elevation.

_____ 10. Exterior walls are either brick veneer or _____″ lap siding.

_____ 11. The O window has a(n) _____ top.

_____ 12. The roof has a total of _____ skylights.

T   F   13. Front entry steps contain a landing.

T   F   14. Deck steps are attached to the rear wall.

_____ 15. The front entry has a(n) _____-panel, insulated metal door.

T   F   16. All concrete foundation footings are at the same level.

_____   **17.** Regarding the deck, _____.
        A. four 12″ diameter concrete piers are shown
        B. stairs are straight-run with two handrails
        C. stairs are located on the West end of the deck
        D. none of the above

T   F   **18.** A combination of gable and shed roofs are shown for the Stewart Residence.

T   F   **19.** Glass blocks are shown in the North and East Elevations.

_____   **20.** The basement has a clear ceiling height of _____.

_____   **21.** B windows are 4′-9″ × _____.

_____   **22.** The front entry steps are made of _____.

_____   **23.** Exterior wall finish above the garage roof is _____.

_____   **24.** Each garage door is 9′-0″ wide × _____ high.

_____   **25.** The K skylight is visible in all elevation views except the _____ Elevation.

# Exam—Details

*Review Questions*

Name _____ Date _____

_____ 1. Elevation Detail 2/5 shows the _____ wall of the kitchen.

_____ 2. Kitchen base cabinets contain a total of _____ drawers.

_____ 3. The refrigerator cabinet has an opening _____ " wide.

T   F   **4.** Wall countertops have a 4" backsplash.

T   F   **5.** Soffits over the wall cabinets are 14" deep.

T   F   **6.** A full glass, insulated metal door leads to the deck.

_____ 7. Regarding the wall cabinets, _____.
   - A. two 21" × 30" cabinets are required
   - B. three 33" × 30" cabinets are required
   - C. all wall cabinets are 12" deep
   - D. none of the above

_____ 8. Regarding the base cabinets, _____.
   - A. the drawer base is 18" wide
   - B. all base cabinets are 36" high
   - C. both A and B
   - D. neither A nor B

T   F   **9.** Elevation 3/5 shows details of the oven cabinet.

_____ 10. Kitchen cabinet elevation details are drawn to the scale of _____.

_____ 11. The Kitchen Plan is drawn to the scale of _____.

_____ 12. Regarding the Kitchen Plan, _____.
   - A. the panned ceiling is shown with dashed lines
   - B. six details are referenced
   - C. both A and B
   - D. neither A nor B

_____ 13. Regarding the sectional view detail of the skylight, _____.
   - A. the skylight is 3'-0" in diameter
   - B. 6" batt insulation surrounds the skylight shaft
   - C. the scale is ⅜" = 1'-0"
   - D. galvanized sheet metal is used as flashing

T   F   **14.** The broom pantry is placed between the lazy Susan pantry and the cooktop.

_____ 15. The eating bar is _____ " lower than the island counter.

_____   **16.** All wall cabinets, except for the hood cabinet, are _____ high.

_____   **17.** The soffit over the broom pantry is _____ deep.

T     F   **18.** The microwave is placed in the oven cabinet above the oven.

_____   **19.** Vertical clearance between the base cabinet countertops and wall cabinets is _____.

T     F   **20.** The sink base cabinet is 36″ wide.

_____   **21.** Soffits over the wall cabinets are _____ high.

T     F   **22.** The lazy Susan pantry is shown in Elevation Detail 1/5.

_____   **23.** Regarding the sectional view detail of the skylight, _____.
     A. panned ceiling framing is 2 × 4s
     B. the cutting plane for the section is shown on the Kitchen Plan
     C. both A and B
     D. neither A nor B

_____   **24.** The opening for the oven is _____″ wide.

T     F   **25.** The vertical clearance between the countertops and wall cabinets is 1′-6″.

# Exam—Sectional Views
*Review Questions*

Name _____ Date _____

_____ 1. The cutting plane for Section 1/6 is shown on Sheets _____.

_____ 2. Sectional views are drawn to the scale of _____.

T   F 3. Sectional views are shown on Sheet 6 of 7.

T   F 4. Roof slope for the Stewart Residence is 5 in 12.

_____ 5. The finish floor level of the den is _____ below the finish floor level of the living room.

_____ 6. Floor-to-ceiling height of the dining room is _____.

T   F 7. The workshop is directly below the master bedroom.

_____ 8. Section 1/6 is a(n) _____ section.

_____ 9. Section 2/6 is a(n) _____ section.

_____ 10. Regarding the chimney, _____.
    A. face brick is used in the future family room and the living room
    B. native stone is used in the living room
    C. both A and B
    D. neither A nor B

_____ 11. Regarding the workshop, _____.
    A. two solid-core doors are shown
    B. six-panel, exterior doors lead to the backyard
    C. both A and B
    D. neither A nor B

T   F 12. Door 8 is to the master bedroom.

_____ 13. The finish floor level of the low storage area is _____ below the finish floor level of the future family room.

T   F 14. North and South walls of the living room are 10'-0" high.

T   F 15. The living room contains a vaulted ceiling.

T   F 16. Sectional views of the Stewart Residence were drawn by PLH.

_____ 17. The stairsteps from the garage to the lower level contain _____ risers.

_____ 18. A(n) _____ beam in the garage supports floor joists above.

T   F 19. The floor-to-ceiling height of the future family room is 7'-9½".

T   F 20. Two casement windows are shown in the sectional view of the dining room.

_____ **21.** Regarding the fireplaces, one is located on the _____.
        A. South wall of the living room
        B. East wall of the future family room
        C. both A and B
        D. neither A nor B

_____ **22.** Closet doors for Bedroom 3 are _____ doors.

T    F    **23.** The size for dining room windows is 4'-9" × 5'-0⅜".

T    F    **24.** The circletop window in the living room contains six pie-shaped lights.

T    F    **25.** The floor-to-ceiling height in all bedrooms is 8'-1⅛".

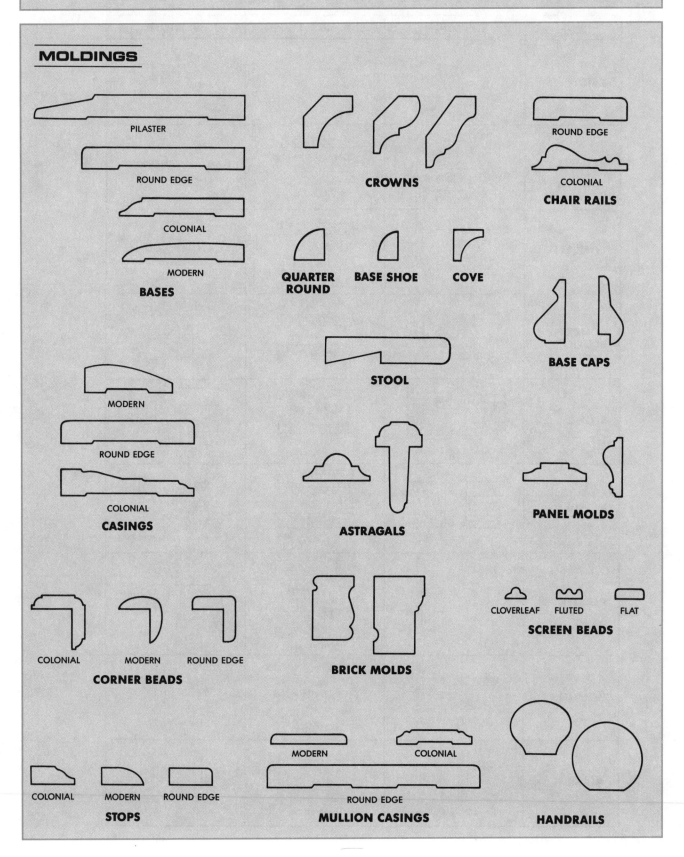

**MOLDINGS**

PILASTER

ROUND EDGE

COLONIAL

MODERN

**BASES**

**CROWNS**

QUARTER ROUND

BASE SHOE

COVE

ROUND EDGE

COLONIAL

**CHAIR RAILS**

**STOOL**

**BASE CAPS**

MODERN

ROUND EDGE

COLONIAL

**CASINGS**

**ASTRAGALS**

**PANEL MOLDS**

COLONIAL    MODERN    ROUND EDGE

**CORNER BEADS**

**BRICK MOLDS**

CLOVERLEAF    FLUTED    FLAT

**SCREEN BEADS**

COLONIAL    MODERN    ROUND EDGE

**STOPS**

MODERN    COLONIAL

ROUND EDGE

**MULLION CASINGS**

**HANDRAILS**

191

# ARCHITECTURAL SYMBOLS

| | ELEVATION | PLAN | SECTION |
|---|---|---|---|
| EARTH | | | |
| BRICK | BRICK — WITH NOTE INDICATING TYPE OF BRICK (COMMON, FACE, ETC.) | COMMON OR FACE / FIREBRICK | SAME AS PLAN VIEWS |
| CONCRETE | | LIGHTWEIGHT / STRUCTURAL | SAME AS PLAN VIEWS |
| CONCRETE BLOCK | | OR | |
| STONE | CUT STONE   RUBBLE | CUT STONE   RUBBLE / CAST STONE (CONCRETE) | CUT STONE / CAST STONE (CONCRETE)  RUBBLE OR CUT STONE |
| WOOD | SIDING   PANEL | WOOD STUD / REMODELING / DISPLAY | ROUGH MEMBERS   FINISHED MEMBERS |
| PLASTER | | WOOD STUD, LATH, AND PLASTER / METAL LATH AND PLASTER / SOLID PLASTER | LATH AND PLASTER |
| ROOFING | SHINGLES | SAME AS ELEVATION | |
| GLASS | OR / GLASS BLOCK | GLASS / GLASS BLOCK | SMALL SCALE   LARGE SCALE |

## ARCHITECTURAL SYMBOLS (continued)

| | ELEVATION | PLAN | SECTION |
|---|---|---|---|
| FACING TILE | CERAMIC TILE | FLOOR TILE | CERAMIC TILE LARGE SCALE / CERAMIC TILE SMALL SCALE |
| STRUCTURAL CLAY TILE | | | SAME AS PLAN VIEW |
| INSULATION | | LOOSE FILL OR BATTS / RIGID / SPRAY FOAM | SAME AS PLAN VIEWS |
| SHEET METAL FLASHING | | OCCASIONALLY INDICATED BY NOTE | |
| METALS OTHER THAN FLASHING | INDICATED BY NOTE OR DRAWN TO SCALE | SAME AS ELEVATION | STEEL / CAST IRON / ALUM / BRONZE OR BRASS — SMALL SCALE |
| STRUCTURAL STEEL | INDICATED BY NOTE OR DRAWN TO SCALE | OR | REBARS / SMALL SCALE / LARGE SCALE / L-ANGLES, S-BEAMS, ETC. |

## PLOT PLAN SYMBOLS

| | | | |
|---|---|---|---|
| N — NORTH | FIRE HYDRANT | WALK | E OR — ELECTRIC SERVICE |
| POINT OF BEGINNING (POB) | MAILBOX | IMPROVED ROAD | G OR — NATURAL GAS LINE |
| UTILITY METER OR VALVE | MANHOLE | UNIMPROVED ROAD | W OR — WATER LINE |
| POWER POLE AND GUY | TREE | BUILDING LINE | T OR — TELEPHONE LINE |
| LIGHT STANDARD | BUSH | PROPERTY LINE | NATURAL GRADE |
| TRAFFIC SIGNAL | HEDGE ROW | PROPERTY LINE | FINISH GRADE |
| STREET SIGN | FENCE | TOWNSHIP LINE | + XX.00' EXISTING ELEVATION |

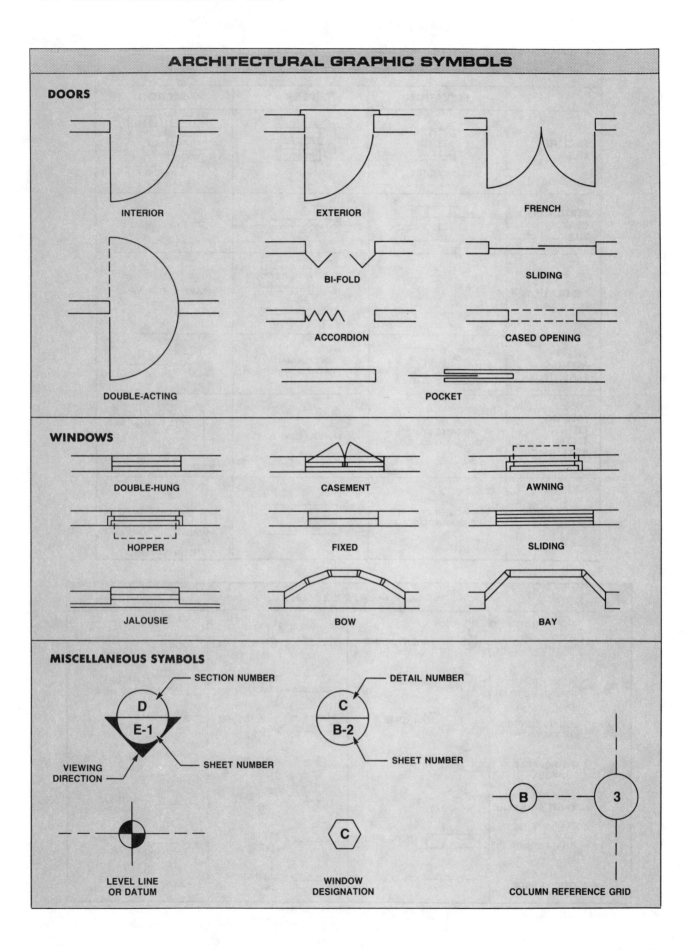

# ARCHITECTURAL GRAPHIC SYMBOLS

**DOORS**

INTERIOR

EXTERIOR

FRENCH

DOUBLE-ACTING

BI-FOLD

SLIDING

ACCORDION

CASED OPENING

POCKET

**WINDOWS**

DOUBLE-HUNG

CASEMENT

AWNING

HOPPER

FIXED

SLIDING

JALOUSIE

BOW

BAY

**MISCELLANEOUS SYMBOLS**

SECTION NUMBER

D
E-1

VIEWING
DIRECTION

SHEET NUMBER

DETAIL NUMBER

C
B-2

SHEET NUMBER

LEVEL LINE
OR DATUM

WINDOW
DESIGNATION

B --- 3

COLUMN REFERENCE GRID

# ELECTRICAL SYMBOLS

## LIGHTING OUTLETS

CEILING, WALL

OUTLET BOX AND INCANDESCENT LIGHTING FIXTURE

INCANDESCENT TRACK LIGHTING

BLANKED OUTLET

DROP CORD

EXIT LIGHT AND OUTLET BOX. SHADED AREAS DENOTE FACES.

OUTDOOR POLE-MOUNTED FIXTURES

JUNCTION BOX

LAMPHOLDER WITH PULL SWITCH

MULTIPLE FLOODLIGHT ASSEMBLY

EMERGENCY BATTERY PACK WITH CHARGER

INDIVIDUAL FLUORESCENT FIXTURE

OUTLET BOX AND FLUORESCENT LIGHTING TRACK FIXTURE

CONTINUOUS FLUORESCENT FIXTURE

SURFACE-MOUNTED FLUORESCENT FIXTURE

### PANELBOARDS

FLUSH-MOUNTED PANELBOARD AND CABINET

SURFACE-MOUNTED PANELBOARD AND CABINET

## CONVENIENCE OUTLETS

SINGLE RECEPTACLE OUTLET

DUPLEX RECEPTACLE OUTLET

TRIPLEX RECEPTACLE OUTLET

SPLIT-WIRED DUPLEX RECEPTACLE OUTLET

SPLIT-WIRED TRIPLEX RECEPTACLE OUTLET

SINGLE SPECIAL-PURPOSE RECEPTACLE OUTLET

DUPLEX SPECIAL-PURPOSE RECEPTACLE OUTLET

RANGE OUTLET

SPECIAL-PURPOSE CONNECTION

CLOSED-CIRCUIT TELEVISION CAMERA

CLOCK HANGER RECEPTACLE

FAN HANGER RECEPTACLE

FLOOR SINGLE RECEPTACLE OUTLET

FLOOR DUPLEX RECEPTACLE OUTLET

FLOOR SPECIAL-PURPOSE OUTLET

UNDERFLOOR DUCT AND JUNCTION BOX FOR TRIPLE, DOUBLE, OR SINGLE DUCT SYSTEM AS INDICATED BY NUMBER OF PARALLEL LINES

### BUSDUCTS AND WIREWAYS

SERVICE, FEEDER, OR PLUG-IN BUSWAY

CABLE THROUGH LADDER OR CHANNEL

WIREWAY

## SWITCH OUTLETS

SINGLE-POLE SWITCH

DOUBLE-POLE SWITCH

THREE-WAY SWITCH

FOUR-WAY SWITCH

AUTOMATIC DOOR SWITCH

KEY-OPERATED SWITCH

CIRCUIT BREAKER

WEATHERPROOF CIRCUIT BREAKER

DIMMER

REMOTE CONTROL SWITCH

WEATHERPROOF SWITCH

FUSED SWITCH

WEATHERPROOF FUSED SWITCH

TIME SWITCH

CEILING PULL SWITCH

SWITCH AND SINGLE RECEPTACLE

SWITCH AND DOUBLE RECEPTACLE

ANY STANDARD SYMBOL WITH THE ADDITION OF A LOWERCASE SUBSCRIPT LETTER MAY BE USED TO DESIGNATE A VARIATION IN STANDARD EQUIPMENT.

# ELECTRICAL SYMBOLS [continued]

## COMMERCIAL AND INDUSTRIAL SYSTEMS

PAGING SYSTEM DEVICE

FIRE ALARM SYSTEM DEVICE

COMPUTER DATA SYSTEM DEVICE

PRIVATE TELEPHONE SYSTEM DEVICE

SOUND SYSTEM

FIRE ALARM CONTROL PANEL — FACP

## SIGNALING SYSTEM OUTLETS FOR RESIDENTIAL SYSTEMS

PUSH BUTTON

BUZZER

BELL

BELL AND BUZZER COMBINATION

COMPUTER DATA OUTLET
*or telephone outlet*

BELL RINGING TRANSFORMER — BT

ELECTRIC DOOR OPENER — D

CHIME — CH

TELEVISION OUTLET — TV

THERMOSTAT — T

## UNDERGROUND ELECTRICAL DISTRIBUTION OR ELECTRICAL LIGHTING SYSTEMS

MANHOLE — M

HANDHOLE — H

TRANSFORMER-MANHOLE OR VAULT — TM

TRANSFORMER PAD — TP

UNDERGROUND DIRECT BURIAL CABLE

UNDERGROUND DUCT LINE

STREET LIGHT STANDARD FED FROM UNDERGROUND CIRCUIT

## ABOVE-GROUND ELECTRICAL DISTRIBUTION OR LIGHTING SYSTEMS

POLE

STREET LIGHT AND BRACKET

PRIMARY CIRCUIT

SECONDARY CIRCUIT

DOWN GUY

HEAD GUY

SIDEWALK GUY

SERVICE WEATHERHEAD

## PANEL CIRCUITS AND MISCELLANEOUS

LIGHTING PANEL

POWER PANEL

WIRING—CONCEALED IN CEILING OR WALL

WIRING—CONCEALED IN FLOOR

WIRING EXPOSED

HOME RUN TO PANEL BOARD.
Indicate number of circuits by number of arrows. Any circuit without such designation indicates a two-wire circuit. For a greater number of wires indicate as follows:
—/// (3 wires) —/// (4 wires), etc.

FEEDERS
Use heavy lines and designate by number corresponding to listing in feeder schedule.

WIRING TURNED UP

WIRING TURNED DOWN

GENERATOR — G

MOTOR — M

INSTRUMENT (SPECIFY) — I

TRANSFORMER — T

CONTROLLER

EXTERNALLY-OPERATED DISCONNECT SWITCH

PULL BOX

# PLUMBING SYMBOLS

## FIXTURES

- STANDARD BATHTUB
- OVAL BATHTUB
- WHIRLPOOL BATH
- SHOWER STALL
- SHOWER HEAD
- TANK-TYPE WATER CLOSET
- WALL-MOUNTED WATER CLOSET
- FLOOR-MOUNTED WATER CLOSET
- LOW-PROFILE WATER CLOSET
- BIDET
- WALL-MOUNTED URINAL
- FLOOR-MOUNTED URINAL
- TROUGH-TYPE URINAL
- WALL-MOUNTED LAVATORY
- PEDESTAL LAVATORY
- BUILT-IN LAVATORY
- WHEELCHAIR LAVATORY
- CORNER LAVATORY
- FLOOR DRAIN
- FLOOR SINK

## FIXTURES (continued)

- LAUNDRY TRAY
- BUILT-IN SINK
- DOUBLE OR TRIPLE BUILT-IN SINK
- COMMERCIAL KITCHEN SINK
- SERVICE SINK
- CLINIC SERVICE SINK
- FLOOR-MOUNTED SERVICE SINK
- DRINKING FOUNTAIN
- WATER COOLER
- HOT WATER TANK
- WATER HEATER
- METER
- HOSE BIBB
- GAS OUTLET
- GREASE SEPARATOR
- GARAGE DRAIN
- FLOOR DRAIN WITH BACKWATER VALVE

## PIPING

- SOIL, WASTE, OR LEADER—ABOVE GRADE
- SOIL, WASTE, OR LEADER—BELOW GRADE
- VENT
- COMBINATION WASTE AND VENT — SV —
- STORM DRAIN — SD —
- COLD WATER

## PIPING (continued)

- CHILLED DRINKING WATER SUPPLY — DWS —
- CHILLED DRINKING WATER RETURN — DWR —
- HOT WATER
- HOT WATER RETURN
- SANITIZING HOT WATER SUPPLY (180°F)
- SANITIZING HOT WATER RETURN (180°F)
- DRY STANDPIPE — DSP —
- COMBINATION STANDPIPE — CSP —
- MAIN SUPPLIES SPRINKLER — S —
- BRANCH AND HEAD SPRINKLER
- GAS—LOW PRESSURE — G — G —
- GAS—MEDIUM PRESSURE — MG —
- GAS—HIGH PRESSURE — HG —
- COMPRESSED AIR — A —
- OXYGEN — O —
- NITROGEN — N —
- HYDROGEN — H —
- HELIUM — HE —
- ARGON — AR —
- LIQUID PETROLEUM GAS — LPG —
- INDUSTRIAL WASTE — INW —
- CAST IRON — CI —
- CULVERT PIPE — CP —
- CLAY TILE — CT —
- DUCTILE IRON — DI —
- REINFORCED CONCRETE — RCP —
- DRAIN—OPEN TILE OR AGRICULTURAL TILE

# PIPE FITTING AND VALVE SYMBOLS

| | FLANGED | SCREWED | BELL & SPIGOT | | FLANGED | SCREWED | BELL & SPIGOT | | FLANGED | SCREWED | BELL & SPIGOT |
|---|---|---|---|---|---|---|---|---|---|---|---|
| BUSHING | | | | REDUCING FLANGE | | | | AUTOMATIC BY-PASS VALVE | | | |
| CAP | | | | BULL PLUG | | | | AUTOMATIC REDUCING VALVE | | | |
| REDUCING CROSS | | | | PIPE PLUG | | | | STRAIGHT CHECK VALVE | | | |
| STRAIGHT-SIZE CROSS | | | | CONCENTRIC REDUCER | | | | COCK | | | |
| CROSSOVER | | | | ECCENTRIC REDUCER | | | | DIAPHRAGM VALVE | | | |
| 45° ELBOW | | | | SLEEVE | | | | FLOAT VALVE | | | |
| 90° ELBOW | | | | STRAIGHT-SIZE TEE | | | | GATE VALVE | | | |
| ELBOW—TURNED DOWN | | | | TEE—OUTLET UP | | | | MOTOR-OPERATED GATE VALVE | | | |
| ELBOW—TURNED UP | | | | TEE—OUTLET DOWN | | | | GLOBE VALVE | | | |
| BASE ELBOW | | | | DOUBLE-SWEEP TEE | | | | MOTOR-OPERATED GLOBE VALVE | | | |
| DOUBLE-BRANCH ELBOW | | | | REDUCING TEE | | | | ANGLE HOSE VALVE | | | |
| LONG-RADIUS ELBOW | | | | SINGLE-SWEEP TEE | | | | GATE VALVE | | | |
| REDUCING ELBOW | | | | SIDE OUTLET TEE—OUTLET DOWN | | | | GLOBE VALVE | | | |
| SIDE OUTLET ELBOW—OUTLET DOWN | | | | SIDE OUTLET TEE—OUTLET UP | | | | LOCKSHIELD VALVE | | | |
| SIDE OUTLET ELBOW—OUTLET UP | | | | UNION | | | | QUICK-OPENING VALVE | | | |
| STREET ELBOW | | | | ANGLE CHECK VALVE | | | | SAFETY VALVE | | | |
| CONNECTING PIPE JOINT | | | | ANGLE GATE VALVE—ELEVATION | | | | | | | |
| EXPANSION JOINT | | | | ANGLE GATE VALVE—PLAN | | | | GOVERNOR-OPERATED AUTOMATIC VALVE | | | |
| LATERAL | | | | ANGLE GLOBE VALVE—ELEVATION | | | | | | | |
| ORIFICE FLANGE | | | | ANGLE GLOBE VALVE—PLAN | | | | | | | |

*The American Society of Mechanical Engineers*

# HVAC SYMBOLS

## EQUIPMENT SYMBOLS

|  | |
|---|---|
| EXPOSED RADIATOR | |
| RECESSED RADIATOR | |
| FLUSH ENCLOSED RADIATOR | |
| PROJECTING ENCLOSED RADIATOR | |
| UNIT HEATER (PROPELLER)—PLAN | |
| UNIT HEATER (CENTRIFUGAL)—PLAN | |
| UNIT VENTILATOR—PLAN | |
| STEAM | |
| DUPLEX STRAINER | |
| PRESSURE REDUCING VALVE | |
| AIR LINE VALVE | |
| STRAINER | |
| THERMOMETER | |
| PRESSURE GAUGE AND COCK | |
| RELIEF VALVE | |
| AUTOMATIC 3-WAY VALVE | |
| AUTOMATIC 2-WAY VALVE | |
| SOLENOID VALVE | |

## DUCTWORK

|  |  |
|---|---|
| DUCT (1ST FIGURE, WIDTH; 2ND FIGURE, DEPTH) | 12 × 20 |
| DIRECTION OF FLOW | |
| FLEXIBLE CONNECTION | |
| DUCTWORK WITH ACOUSTICAL LINING | |
| FIRE DAMPER WITH ACCESS DOOR | FD  AD |
| MANUAL VOLUME DAMPER | VD |
| AUTOMATIC VOLUME DAMPER | |
| EXHAUST, RETURN OR OUTSIDE AIR DUCT—SECTION | 20 × 12 |
| SUPPLY DUCT—SECTION | 20 × 12 |
| CEILING DIFFUSER SUPPLY OUTLET | 20" DIA. CD 1000 CFM |
| CEILING DIFFUSER SUPPLY OUTLET | 20 × 12 CD 700 CFM |
| LINEAR DIFFUSER | 96 × 6-LD 400 CFM |
| FLOOR REGISTER | 20 × 12 FR 700 CFM |
| TURNING VANES | |
| FAN AND MOTOR WITH BELT GUARD | |
| LOUVER OPENING | 20 × 12-L 700 CFM |

## HEATING PIPING

| Symbol Name | Abbr. |
|---|---|
| HIGH PRESSURE STEAM | HPS |
| MEDIUM PRESSURE STEAM | MPS |
| LOW PRESSURE STEAM | LPS |
| HIGH PRESSURE RETURN | HPR |
| MEDIUM PRESSURE RETURN | MPR |
| LOW PRESSURE RETURN | LPR |
| BOILER BLOW OFF | BD |
| CONDENSATE OR VACUUM PUMP DISCHARGE | VPD |
| FEEDWATER PUMP DISCHARGE | PPD |
| MAKE UP WATER | MU |
| AIR RELIEF LINE | V |
| FUEL OIL SUCTION | FOS |
| FUEL OIL RETURN | FOR |
| FUEL OIL VENT | FOV |
| COMPRESSED AIR | A |
| HOT WATER HEATING SUPPLY | HW |
| HOT WATER HEATING RETURN | HWR |

## AIR CONDITIONING PIPING

| Symbol Name | Abbr. |
|---|---|
| REFRIGERANT LIQUID | RL |
| REFRIGERANT DISCHARGE | RD |
| REFRIGERANT SUCTION | RS |
| CONDENSER WATER SUPPLY | CWS |
| CONDENSER WATER RETURN | CWR |
| CHILLED WATER SUPPLY | CHWS |
| CHILLED WATER RETURN | CHWR |
| MAKE UP WATER | MU |
| HUMIDIFICATION LINE | H |
| DRAIN | D |

## ALPHABET OF LINES

| LINE | REPRESENTATION | | USE |
|------|----------------|--|-----|
| Object line | ——————————— |  THICK | Defines shape. |
| Hidden line | – – – – – – – – – – ⅛″ 1/32″ | 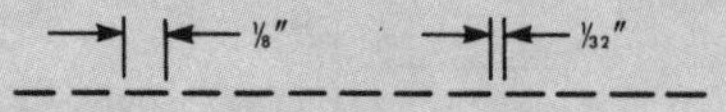 MEDIUM | Shows hidden features, future or existing construction to be removed. |
| Centerline | ¾″–1½″ ⅛″ 1/16″ | 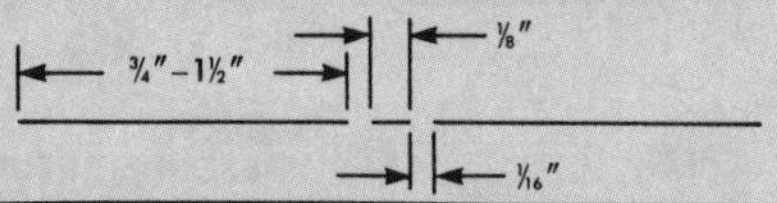 THIN | Locates centerpoints of arcs and circles, exterior elevation lines, and projections. |
| Dimension line | 4′-0″ SLASH<br>4′-0″ DOT<br>4′-0″ ARROWHEAD | 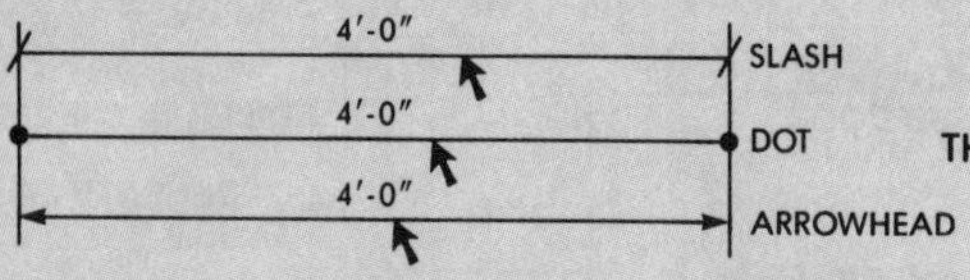 THIN | Shows size or location. |
| Extension line | | THIN | Defines size or location. |
| Leader | |  THIN | Indicates specific features. |
| Long break line | ¾″–1½″ | 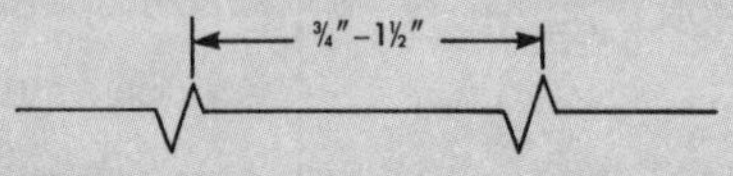 THIN | Indicates long breaks. |
| Short break line | |  MEDIUM | Indicates short breaks. |
| Cutting plane | ¾″–1½″ 1/16″ ⅛″ | 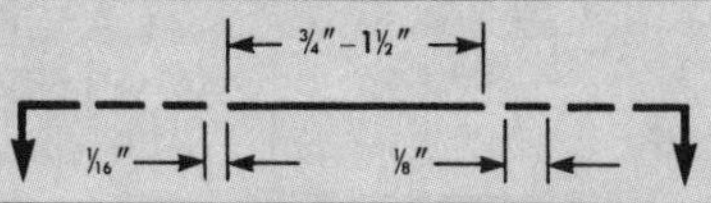 THICK | Shows internal features. |
| Section line | 1/16″ | 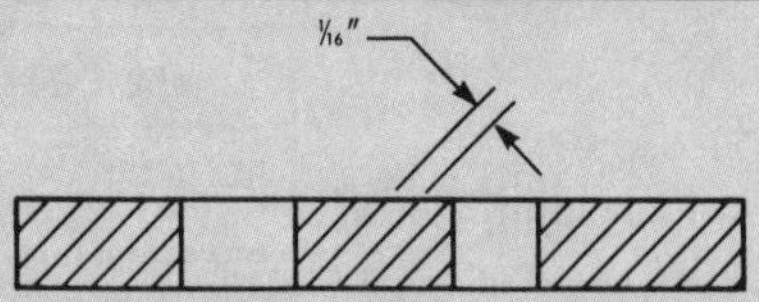 MEDIUM | Indicates internal features. |
| Phantom line | ¾″–1½″ ⅛″ 1/16″ | 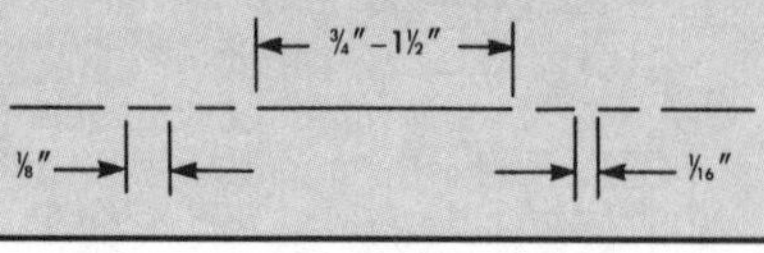 THIN | Indicates movement, property, and boundary lines. |

## CONCRETE BLOCK AND BRICK

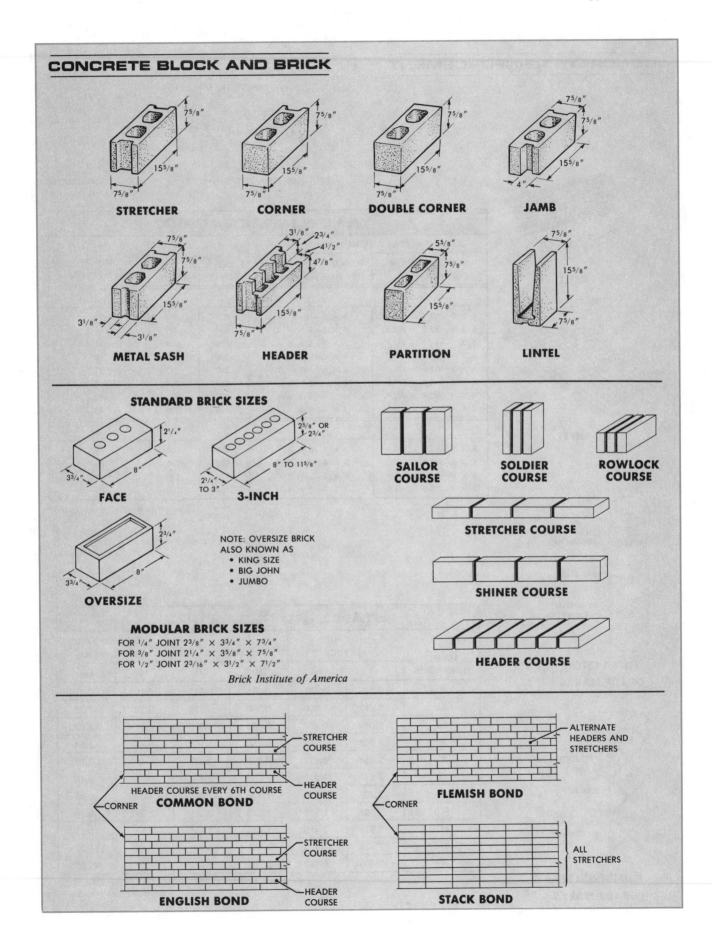

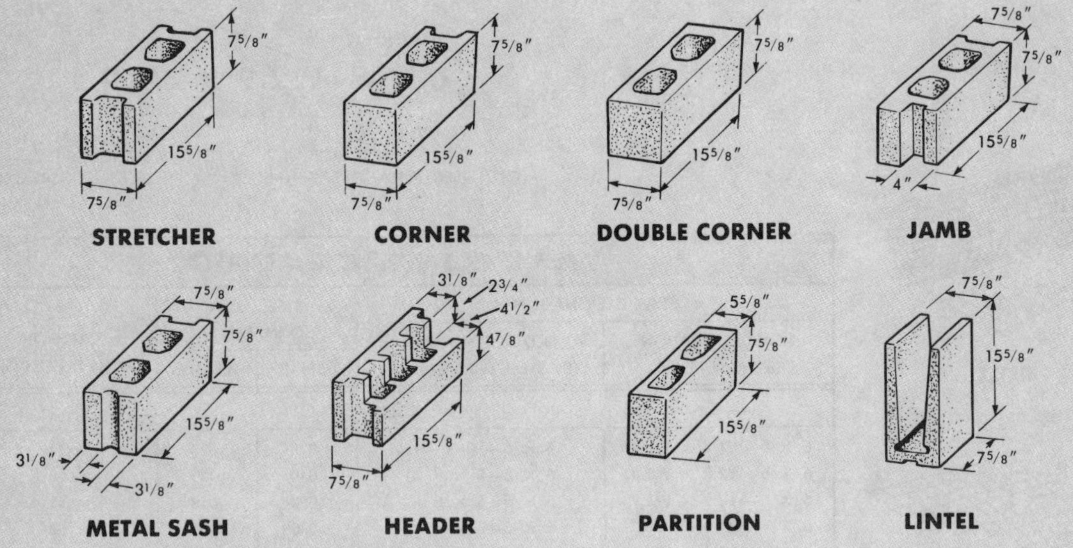
7 5/8"
15 5/8"
7 5/8"
STRETCHER
7 5/8"
15 5/8"
7 5/8"
CORNER
7 5/8"
15 5/8"
7 5/8"
DOUBLE CORNER
7 5/8"
7 5/8"
15 5/8"
4"
JAMB
7 5/8"
7 5/8"
15 5/8"
3 1/8"
3 1/8"
METAL SASH
3 1/8"
2 3/4"
4 1/2"
4 7/8"
15 5/8"
7 5/8"
HEADER
5 5/8"
7 5/8"
15 5/8"
PARTITION
7 5/8"
15 5/8"
7 5/8"
LINTEL

**STANDARD BRICK SIZES**

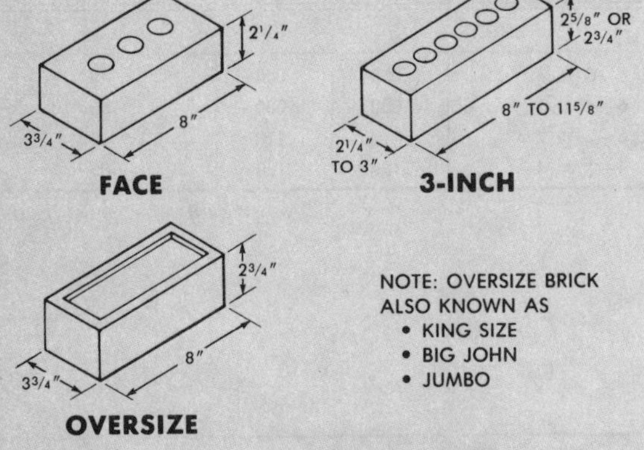
2 1/4"
8"
3 3/4"
FACE
2 5/8" OR 2 3/4"
8" TO 11 5/8"
2 1/4" TO 3"
3-INCH
2 3/4"
8"
3 3/4"
OVERSIZE

NOTE: OVERSIZE BRICK ALSO KNOWN AS
- KING SIZE
- BIG JOHN
- JUMBO

**MODULAR BRICK SIZES**

FOR 1/4" JOINT 2 3/8" × 3 3/4" × 7 3/4"
FOR 3/8" JOINT 2 1/4" × 3 5/8" × 7 5/8"
FOR 1/2" JOINT 2 3/16" × 3 1/2" × 7 1/2"

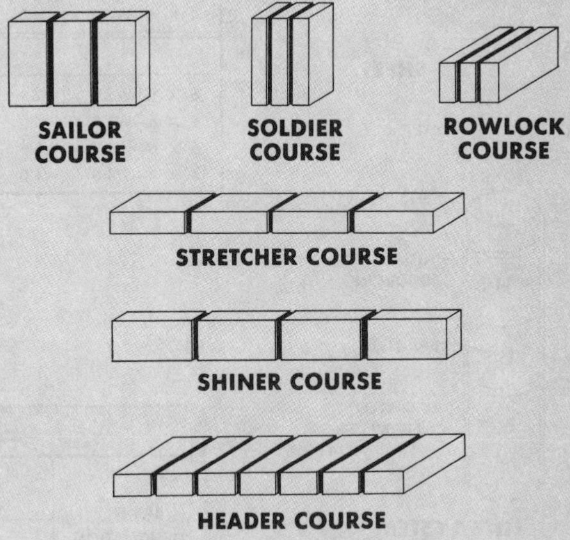
SAILOR COURSE
SOLDIER COURSE
ROWLOCK COURSE
STRETCHER COURSE
SHINER COURSE
HEADER COURSE

*Brick Institute of America*

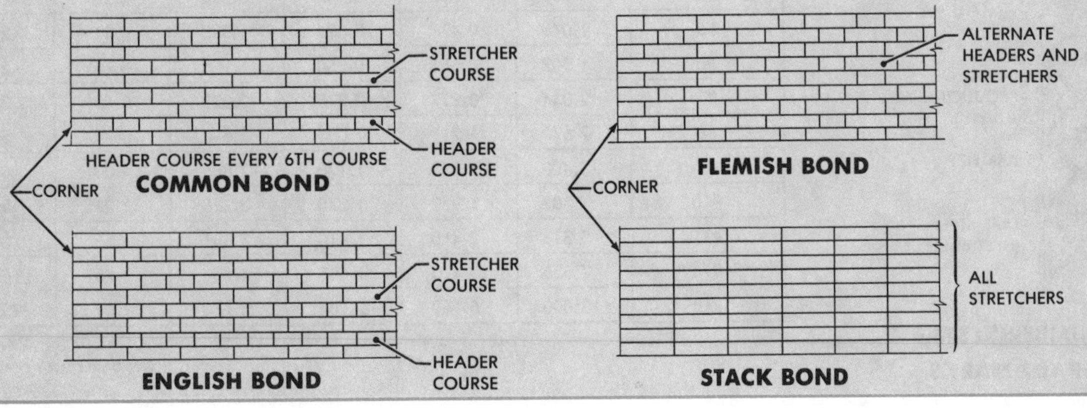
STRETCHER COURSE
HEADER COURSE
HEADER COURSE EVERY 6TH COURSE
CORNER
COMMON BOND
ALTERNATE HEADERS AND STRETCHERS
CORNER
FLEMISH BOND
STRETCHER COURSE
HEADER COURSE
ENGLISH BOND
ALL STRETCHERS
STACK BOND

## CONCRETE REINFORCEMENT

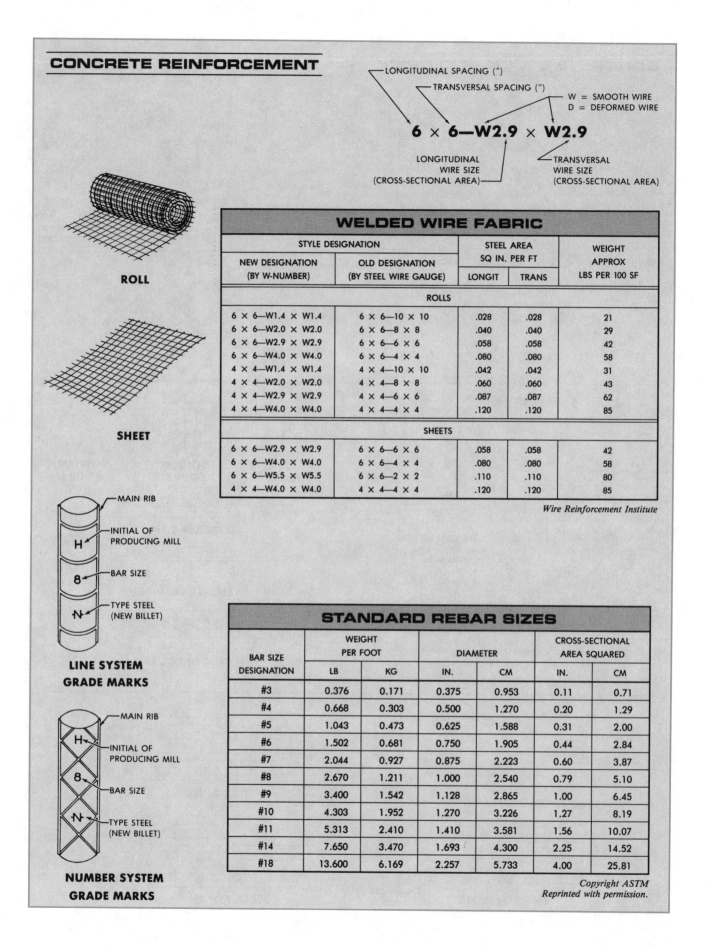

LONGITUDINAL SPACING (")
TRANSVERSAL SPACING (")
W = SMOOTH WIRE
D = DEFORMED WIRE

$$6 \times 6 - W2.9 \times W2.9$$

LONGITUDINAL WIRE SIZE (CROSS-SECTIONAL AREA)

TRANSVERSAL WIRE SIZE (CROSS-SECTIONAL AREA)

ROLL

SHEET

MAIN RIB
INITIAL OF PRODUCING MILL
BAR SIZE
TYPE STEEL (NEW BILLET)

**LINE SYSTEM GRADE MARKS**

MAIN RIB
INITIAL OF PRODUCING MILL
BAR SIZE
TYPE STEEL (NEW BILLET)

**NUMBER SYSTEM GRADE MARKS**

### WELDED WIRE FABRIC

| STYLE DESIGNATION | | STEEL AREA SQ IN. PER FT | | WEIGHT APPROX LBS PER 100 SF |
|---|---|---|---|---|
| NEW DESIGNATION (BY W-NUMBER) | OLD DESIGNATION (BY STEEL WIRE GAUGE) | LONGIT | TRANS | |
| ROLLS | | | | |
| 6 × 6—W1.4 × W1.4 | 6 × 6—10 × 10 | .028 | .028 | 21 |
| 6 × 6—W2.0 × W2.0 | 6 × 6—8 × 8 | .040 | .040 | 29 |
| 6 × 6—W2.9 × W2.9 | 6 × 6—6 × 6 | .058 | .058 | 42 |
| 6 × 6—W4.0 × W4.0 | 6 × 6—4 × 4 | .080 | .080 | 58 |
| 4 × 4—W1.4 × W1.4 | 4 × 4—10 × 10 | .042 | .042 | 31 |
| 4 × 4—W2.0 × W2.0 | 4 × 4—8 × 8 | .060 | .060 | 43 |
| 4 × 4—W2.9 × W2.9 | 4 × 4—6 × 6 | .087 | .087 | 62 |
| 4 × 4—W4.0 × W4.0 | 4 × 4—4 × 4 | .120 | .120 | 85 |
| SHEETS | | | | |
| 6 × 6—W2.9 × W2.9 | 6 × 6—6 × 6 | .058 | .058 | 42 |
| 6 × 6—W4.0 × W4.0 | 6 × 6—4 × 4 | .080 | .080 | 58 |
| 6 × 6—W5.5 × W5.5 | 6 × 6—2 × 2 | .110 | .110 | 80 |
| 4 × 4—W4.0 × W4.0 | 4 × 4—4 × 4 | .120 | .120 | 85 |

*Wire Reinforcement Institute*

### STANDARD REBAR SIZES

| BAR SIZE DESIGNATION | WEIGHT PER FOOT | | DIAMETER | | CROSS-SECTIONAL AREA SQUARED | |
|---|---|---|---|---|---|---|
| | LB | KG | IN. | CM | IN. | CM |
| #3 | 0.376 | 0.171 | 0.375 | 0.953 | 0.11 | 0.71 |
| #4 | 0.668 | 0.303 | 0.500 | 1.270 | 0.20 | 1.29 |
| #5 | 1.043 | 0.473 | 0.625 | 1.588 | 0.31 | 2.00 |
| #6 | 1.502 | 0.681 | 0.750 | 1.905 | 0.44 | 2.84 |
| #7 | 2.044 | 0.927 | 0.875 | 2.223 | 0.60 | 3.87 |
| #8 | 2.670 | 1.211 | 1.000 | 2.540 | 0.79 | 5.10 |
| #9 | 3.400 | 1.542 | 1.128 | 2.865 | 1.00 | 6.45 |
| #10 | 4.303 | 1.952 | 1.270 | 3.226 | 1.27 | 8.19 |
| #11 | 5.313 | 2.410 | 1.410 | 3.581 | 1.56 | 10.07 |
| #14 | 7.650 | 3.470 | 1.693 | 4.300 | 2.25 | 14.52 |
| #18 | 13.600 | 6.169 | 2.257 | 5.733 | 4.00 | 25.81 |

*Copyright ASTM*
*Reprinted with permission.*

## STANDARD LUMBER SIZES

| TYPE | THICKNESS | | WIDTH | |
|---|---|---|---|---|
| | Nominal Size | Actual Size | Nominal Size | Actual Size |
| COMMON BOARDS | 1″ | ³/₄″ | 2″<br>4″<br>6″<br>8″<br>10″<br>12″ | 1¹/₂″<br>3¹/₂″<br>5¹/₂″<br>7¹/₄″<br>9¹/₄″<br>11¹/₄″ |
| DIMENSION | 2″ | 1¹/₂″ | 2″<br>4″<br>6″<br>8″<br>10″<br>12″ | 1¹/₂″<br>3¹/₂″<br>5¹/₂″<br>7¹/₄″<br>9¹/₄″<br>11¹/₄″ |
| TIMBERS | 4″<br>6″<br>8″ | 3¹/₂″<br>5¹/₂″<br>7¹/₂″ | 4″<br>6″<br>8″<br>10″ | 3¹/₂″<br>5¹/₂″<br>7¹/₂″<br>9¹/₂″ |
| | 6″ | 5¹/₂″ | 6″<br>8″<br>10″ | 5¹/₂″<br>7¹/₂″<br>9¹/₂″ |
| | 8″ | 7¹/₂″ | 8″<br>10″ | 7¹/₂″<br>9¹/₂″ |

## SPANS

**40# Live Load**
**10# Dead Load**

Design Criteria:
Strength—10 lbs. per sq. ft. dead load plus 40 lbs. per sq. ft. live load.
Deflection—Limited to span in inches divided by 360 for live load only.

**L/360**

| Species or Group | Grade* | Span (feet and inches) | | | | | | | | | | | |
|---|---|---|---|---|---|---|---|---|---|---|---|---|---|
| | | 2 × 6 | | | 2 × 8 | | | 2 × 10 | | | 2 × 12 | | |
| | | 12″ O.C. | 16″ O.C. | 24″ O.C. | 12″ O.C. | 16″ O.C. | 24″ O.C. | 12″ O.C. | 16″ O.C. | 24″ O.C. | 12″ O.C. | 16″ O.C. | 24″ O.C. |
| DOUGLAS FIR-LARCH | 2 | 10-11 | 9-11 | 8-6 | 14-4 | 13-1 | 11-3 | 18-4 | 16-9 | 14-5 | 22-4 | 20-4 | 17-6 |
| | 3 | 9-3 | 8-0 | 6-6 | 12-2 | 10-7 | 8-8 | 15-7 | 13-6 | 11-0 | 18-11 | 16-5 | 13-5 |
| DOUGLAS FIR SOUTH | 2 | 10-0 | 9-1 | 7-11 | 13-2 | 12-0 | 10-6 | 16-9 | 15-3 | 13-4 | 20-5 | 18-7 | 16-3 |
| | 3 | 9-0 | 7-9 | 6-4 | 11-9 | 10-3 | 8-4 | 15-1 | 13-1 | 10-8 | 18-4 | 15-11 | 13-0 |
| HEM-FIR | 2 | 10-3 | 9-4 | 7-7 | 13-6 | 12-3 | 10-0 | 17-3 | 15-8 | 12-10 | 20-11 | 19-1 | 15-7 |
| | 3 | 8-3 | 7-2 | 5-10 | 10-10 | 9-5 | 7-8 | 13-10 | 12-0 | 9-10 | 16-10 | 14-7 | 11-11 |
| MOUNTAIN HEMLOCK-HEM-FIR | 2 | 9-5 | 8-7 | 7-6 | 12-5 | 11-4 | 9-11 | 15-11 | 14-6 | 12-8 | 19-4 | 17-7 | 15-4 |
| | 3 | 8-3 | 7-2 | 5-10 | 10-10 | 9-5 | 7-8 | 13-10 | 12-0 | 9-10 | 16-10 | 14-7 | 11-11 |
| WESTERN HEMLOCK | 2 | 10-3 | 9-4 | 7-11 | 13-6 | 12-3 | 10-6 | 17-3 | 15-8 | 13-4 | 20-11 | 19-1 | 16-3 |
| | 3 | 8-8 | 7-6 | 6-1 | 11-5 | 9-11 | 8-1 | 14-7 | 12-8 | 10-4 | 17-9 | 15-5 | 12-7 |
| ENGLEMANN SPRUCE LODGEPOLE PINE (Englemann Spruce-Alpine Fir) | 2 | 9-5 | 8-7 | 6-11 | 12-5 | 11-2 | 9-1 | 15-11 | 14-3 | 11-7 | 19-4 | 17-3 | 14-2 |
| | 3 | 7-5 | 6-5 | 5-3 | 9-9 | 8-6 | 6-11 | 12-6 | 10-10 | 8-10 | 15-3 | 13-2 | 10-9 |
| LODGEPOLE PINE | 2 | 9-8 | 8-10 | 7-3 | 12-10 | 11-8 | 9-7 | 16-4 | 14-11 | 12-3 | 19-10 | 18-1 | 14-11 |
| | 3 | 7-10 | 6-10 | 5-7 | 10-5 | 9-1 | 7-5 | 13-4 | 11-7 | 9-5 | 16-3 | 14-1 | 11-6 |
| PONDEROSA PINE-LODGEPOLE PINE | 2 | 9-5 | 8-7 | 7-0 | 12-5 | 11-4 | 9-3 | 15-11 | 14-5 | 11-9 | 19-4 | 17-7 | 14-4 |
| | 3 | 7-7 | 6-6 | 5-4 | 10-0 | 8-8 | 7-1 | 12-9 | 11-1 | 9-1 | 15-7 | 13-6 | 11-0 |
| WESTERN CEDARS | 2 | 9-2 | 8-4 | 7-3 | 12-0 | 11-0 | 9-7 | 15-4 | 14-0 | 12-3 | 18-9 | 17-0 | 14-11 |
| | 3 | 7-10 | 6-10 | 7-6 | 10-5 | 9-1 | 7-5 | 13-4 | 11-6 | 9-5 | 16-3 | 14-0 | 11-6 |
| WHITE WOODS (Western Woods) | 2 | 9-2 | 8-4 | 6-10 | 12-0 | 11-0 | 9-0 | 15-5 | 14-0 | 11-6 | 18-9 | 17-0 | 14-0 |
| | 3 | 7-5 | 6-5 | 5-3 | 9-9 | 8-6 | 6-11 | 12-6 | 10-10 | 8-10 | 15-3 | 13-2 | 10-9 |

*Spans were computed for commonly marketed grades. Spans for other grades can be computed utilizing the WWPA Span Computer.

*Western Wood Products Association*

# ABBREVIATIONS

## A

above . . . . . . . . . . . . . . . . . . ABV
access . . . . . . . . . . . . . . . . . . ACS
access panel . . . . . . . . . . . . AP
acoustic . . . . . . . . AC or ACST
acoustical plaster
   ceiling . . . . . . . . . . . . . . APC
acoustical tile . . . . AT. or ACT.
adjacent . . . . . . . . . . . . . . ADJ
adjustable . . . . . . ADJT or ADJ
aggregate . . . . AGG or AGGR
air circulating . . . . . . . . ACIRC
air conditioner . . . . AIR COND
air conditioning . . . . . . . A/C
           or AIR COND
alloy . . . . . . . . . . . . . . . . . ALY
alloy steel . . . . . . . . . ALY STL
alternate . . . . . . . . . . . . . ALTN
alternating current . . . . . . . AC
aluminum . . . . . . . . . . . . . . AL
ambient . . . . . . . . . . . . . . AMB
American National
   Standard . . AMER NATL STD
American National Standards
   Institute . . . . . . . . . . . . ANSI
American Steel Wire
   Gauge . . . . . . . . . . . . ASWG
American Wire Gauge . . AWG
ampere . . . . . . . . . . A or AMP
anchor . . . . . . . . . . . . . . . AHR
anchor bolt . . . . . . . . . . . . . AB
appearance . . . . . . . . . . . APP
apartment . . . . . . . . . . . . . APT.
approximate . . . . . . . . . . APX
          or APPROX
architectural . . . . . . . . . ARCH.
architecture . . . . . . . . . . ARCH.
area . . . . . . . . . . . . . . . . . . . A
area drain . . . . . . . . . . . . . AD
asbestos . . . . . . . . . . . . . ASB
asbestos board . . . . . . . . . AB
as drawn . . . . . . . . . . . . . . AD
asphalt . . . . . . . . . . . . . . ASPH
asphalt roof
   shingles . . . . . . . . . ASPHRS
asphalt tile . . . . . . . . . . . . AT.
automatic . . . . . . . . . . . AUTO.
auxiliary . . . . . . . . . . . . . AUX
avenue . . . . . . . . . . . . . . AVE
azimuth . . . . . . . . . . . . . . . AZ

## B

barrier . . . . . . . . . . . . . . . BARR
barrier, moisture vapor-
   proof . . . . . . . . . . . . . BMVP
barrier, waterproof . . . . . BWP
basement . . . . . . . . . . . BSMT
bathroom . . . . . . . . . . . . . . . B
bathtub . . . . . . . . . . . . . . . BT
batten . . . . . . . . . . . . . . BATT
beam . . . . . . . . . . . . . . . . BM
bearing . . . . . . . . . . . . . . BRG
bearing plate . . . . . . . . . BPL
          or BRG PL
bedroom . . . . . . . . . . . . . . BR

below . . . . . . . . . . . . . . . BLW
bench mark . . . . . . . . . . . BM
beveled wood siding . . . BWS
bituminous . . . . . . . . . . . . BIT.
blocking . . . . . . . . . . . . BLKG
blueprint . . . . . . . . . . . . . . BP
board . . . . . . . . . . . . . . . . BD
board foot . . . . . . BF or BD FT
boiler . . . . . . . . . . . . . . . BLR
bookcase . . . . . . . . . . . . . BC
book shelves . . . . . . . . BK SH
boulevard . . . . . . . . . . . BLVD
boundary . . . . . . . . . . . BDRY
brass . . . . . . . . . . . . . . . . BRS
breaker . . . . . . . . . . . . . BRKR
brick . . . . . . . . . . . . . . . . BRK
British thermal unit . . . . . . Btu
bronze . . . . . . . . . . . . . . BRZ
broom closet . . . . . . . . . . . BC
building . . . . . . . . BLDG or BL
building line . . . . . . . . . . . . BL
built-in . . . . . . . . . . . . . BLTIN
built-up roofing . . . . . . . . BUR

## C

cabinet . . . . . . . . . . . . . CAB.
cable . . . . . . . . . . . . . . . . CA
canopy . . . . . . . . . . . . . CAN.
caulking . . . . . . . CK OR CLKG
cantilever . . . . . . . . . . . CANV
carpenter . . . . . . . . . . CPNTR
cased opening . . . . . . . . . CO
casing . . . . . . . . . . . . . . CSG
cast iron . . . . . . . . . . . . . . CI
cast-iron pipe . . . . . . . . . CIP
cast steel . . . . . . . . . . . . . CS
cast stone . . . . . . CST or CS
catch basin . . . . . . . . . . . CB
caulked joint . . . . . . . . . CLKJ
cavity . . . . . . . . . . . . . . . CAV
ceiling . . . . . . . . . . . . . . CLG
cellar . . . . . . . . . . . . . . . CEL
cement . . . . . . . . . . . . . CEM
cement asbestos
   board . . . . . . . . . CEM AB
cement floor . . . . . . . . . . . CF
cement mortar . . . CEM MORT
center . . . . . . . . . . . . . . CTR
centerline . . . . . . . . . . . . . CL
center matched . . . . . . . . CM
center-to-center . . . . . . C TO C
central . . . . . . . . . . . . . . CTL
ceramic . . . . . . . . . . . . . CER
ceramic tile . . . . . . . . . . . CT
ceramic-tile base . . . . . . CTB
ceramic-to-metal
   (seal) . . . . . . . . . . CERMET
chamfer . . . . CHAM or CHMFR
channel . . . . . . . . . . . . CHAN
check valve . . . . . . . . . . . CV
chimney . . . . . . . . . . . . CHM
chord . . . . . . . . . . . . . . CHD
cinder block . . . . . . . . CINBL
circle . . . . . . . . . . . . . . . CIR
circuit . . . . . . . . . . . . . . . CKT

circuit breaker . . . . . . . . . CB
          or CIR BKR
circuit interrupter . . . . . . . . CI
circumference . . . . . . CRCMF
cleanout . . . . . . . . . . . . . CO
clear glass . . . . . . . . . CL GL
closet . . . . C, CL, CLO, or CLOS
coaxial . . . . . . . . . . . COAX.
cold air . . . . . . . . . . . . . . CA
cold-rolled . . . . . . . . . . . . CR
cold-rolled steel . . . . . . . CRS
cold water . . . . . . . . . . . CW
collar beam . . . . . . . . . COL B
color code . . . . . . . . . . . . CC
combination . . . . . . . . COMB.
combustible . . . . . . . . COMBL
combustion . . . . . . . . COMB.
common . . . . . . . . . . . . COM
composition . . . . . . . . COMP
concrete . . . . . . . . . . . CONC
concrete block . . . . . . . . CCB
       or CONC BLK
concrete floor . . . . . . . . CCF,
  CONC FLR, or CONC FL
concrete masonry unit . . CMU
concrete pipe . . . . . . . . . CP
concrete splash block . . . CSB
condenser . . . . . . . . . . COND
conductor . . . . . . . . . CNDCT
conduit . . . . . . . . . . . . . CND
construction . . . . . . . CONSTR
construction joint . . . . . . . CJ
      or CONSTR JT
continuous . . . . . . . . . CONT
contour . . . . . . . . . . . . . CTR
contract . . . . CONTR or CONT
contractor . . . . . . . . . CONTR
control joint . . . . . . CJ or CLJ
conventional . . . . . . . . CVNTL
copper . . . . . . CPR or COP.
corner . . . . . . . . . . . . . . COR
cornice . . . . . . . . . . . . . COR
corrugate . . . . . . . . . . . CORR
counter . . . . . . . . . . . . CNTR
county . . . . . . . . . . . . . . CO
cubic . . . . . . . . . . . . . . . CU
cubic feet . . . . CFT or CU FT
cubic foot per minute . . . CFM
cubic foot per second . . . . CFS
cubic inch . . . . . . . . . CU IN.
cubic yard . . . . . . . . . CU YD
current . . . . . . . . . . . . . CUR
cutoff . . . . . . . . . . . . . . . CO
cutoff valve . . . . . . . . . . COV
cut out . . . . . . . . . . . . . . CO

## D

damper . . . . . . . . . . . DMPR
datum . . . . . . . . . . . . . . DAT
decibel . . . . . . . . . . . . . . DB
degree . . . . . . . . . . . . . DEG
depth . . . . . . . . . . . . . . . DP
design . . . . . . . . . . . . . DSGN
detail . . . . . . . . . DTL or DET
diagonal . . . . . . . . . . . DIAG

diagram . . . . . . . . . . . . DIAG
dimension . . . . . . . . . . . DIM.
dimmer . . . . . . . DIM. or DMR
dining room . . DR or DNG RM
direct current . . . . . . . . . . DC
direction . . . . . . . . . . . . . DIR
disconnect . . . . . . . . . . DISC.
disconnect switch . . . . . . . DS
dishwasher . . . . . . . . . . . DW
distribution panel . . . . . DPNL
ditto . . . . . . . . . . . . . . . DO.
door . . . . . . . . . . . . . . . . DR
door stop . . . . . . . . . . . DST
door switch . . . . . . . . . . DSW
dormer . . . . . . . . . . . . . DRM
double-acting . . . . . . . . . . DA
      or DBL ACT
double-hung
   window . . . . . . . . . . . DHW
double-pole double-
   throw . . . . . . . . . . . . DPDT
double-pole double-throw
   switch . . . . . . . . . DPDT SW
double-pole single-
   throw . . . . . . . . . . . . DPST
double-pole single-
   throw switch . . . . . DPST SW
double-pole switch . . . DP SW
double-strength glass . . . DSG
down . . . . . . . . . . . . DN or D
downspout . . . . . . . . . . . DS
dozen . . . . . . . . . . . . . . DOZ
drain . . . . . . . . . . . . D or DR
drain tile . . . . . . . . . . . . . DT
drawer . . . . . . . . . . . . . DWR
drawing . . . . . . . . . . . . DWG
dryer . . . . . . . . . . . . . . . . . D
drywall . . . . . . . . . . . . . . DW
dwelling . . . . . . . . . . . DWEL

## E

each . . . . . . . . . . . . . . . . EA
east . . . . . . . . . . . . . . . . . . E
elbow . . . . . . . . . . . . . . ELB
electric or electrical . . . . ELEC
electrical metallic
   tubing . . . . . . . . . . . . EMT
electric operator . . ELECT. OPR
electric panel . . . . . . . . . . EP
electromechanical . . . . ELMCH
elevation . . . . . . . . . . . ELEV
enamel . . . . . . . . . . . ENAM
end-to-end . . . . . . . . . E to E
entrance . . . . . . . . . . . ENTR
equipment . . . . . . . . . . EQPT
equivalent . . . . . . . . . EQUIV
estimate . . . . . . . . . . . . EST
example . . . . . . . . . . . . . EX
excavate . . . . . EXCA or EXC
exchange . . . . . . . . . . EXCH
exhaust . . . . . . . . . . . . EXH
exhaust vent . . . . . . . . EXHV
existing . . . . . . . . . . . . EXST
expanded metal . . . . . . . EM
expansion joint . . . . . . EXP JT

# ABBREVIATIONS

exterior . . . . . . . . . . . . . . . . .EXT
exterior grade . . . . . . .EXT GR

**F**

face brick . . . . . . . . . . . . . .FB
faceplate . . . . . . . . . . . . . . .FP
Fahrenheit . . . . . . . . . . . . . .°F
fiberboard, solid . . . . . . .FBDS
finish . . . . . . . . . .FIN. or FNSH
finish all over . . . . . . . . . .FAO
finish grade . . . . . . . . . . . . .FG
finish one side . . . . . . . . . .F1S
finish two sides . . . . . . . . .F2S
finished floor . . . . . .FIN. FLR,
          FIN. FL, or FNSH FL
firebrick . . . . . . .FBRK or FBCK
fireplace . . . . . . . . . .FPL or FP
fireproof . . . . . . . . .FP or FPRF
fire-resistant . . . . . . . . . .FRES
fixed transom . . . . . . . . . .FTR
fixed window . . . . . .FX WDW
fixture . . . . . . . . .FIX. or FXTR
flashing . . . . . . . . . .FLG or FL
floor . . . . . . . . . . . .FLR or FL
floor drain . . . . . . . . . . . . . .FD
flooring . . . . . . . . . .FLR or FLG
fluorescent . . .FLUR or FLUOR
flush . . . . . . . . . . . . . . . . . . .FL
footing . . . . . . . . . . . . . . . .FTG
foundation . . . . .FND or FDN
frame . . . . . . . . . . . . . . . . . .FR
frostproof hose bibb . . . .FPHB
full scale . . . . . . . . . . . . . .FSC
full size . . . . . . . . . . . . . . . . .FS
furnace . . . . . . . . . . . . . .FURN
furred ceiling . . . . . . . . . . . .FC
furring . . . . . . . . . . . . . . . .FUR
fuse . . . . . . . . . . . . . . . . . . .FU
fuse block . . . . . . . . . . . . . . .FB
fusebox . . . . . . . . . . . . . .FUBX
fuseholder . . . . . . . . . .FUHLR
fusible . . . . . . . . . . . . . . . .FSBL

**G**

gallon . . . . . . . . . . . . . . . .GAL
gallon per hour . . . . . . . .GPH
gallon per minute . . . . . .GPM
galvanized iron . . . . . . . . . .GI
          or GALVI
galvanized steel . . . . . . . . .GS
          or GALVS
garage . . . . . . . . . . . . . . .GAR.
gas . . . . . . . . . . . . . . . . . . . .G
gate valve . . . . . . . . . . . .GTV
gauge . . . . . . . . . . . . . . . .GA
general
    contractor . . . . . .GEN CONT
girder . . . . . . . . . . . . . . . . . .G
glass . . . . . . . . . . . . . . . . . .GL
glass block . . .GLB or GL BL
glaze . . . . . . . . . . . . . . . .GLZ
grade . . . . . . . . . . . . . . . . .GR
grade line . . . . . . . . . . . . . .GL
gravel . . . . . . . . . . . . . . .GVL
grill . . . . . . . . . . . . . . . . . . . .G

gross weight . . . . . . . . .GRWT
ground . . . . . . . . . . . . . . .GRD
grounded (outlet) . . . . . . . .G
ground fault circuit
    interrupter . . . . . . . . . .GFCI
ground fault
    interrupter . . . . . . . . . . .GFI
gypsum . . . . . . . . . . . . . .GYP
gypsum board . . . . . .GYP BD
gypsum-plaster ceiling . . .GPC
gypsum-plaster wall . . . .GPW
gypsum sheathing
    board . . . . . . . . . . . . . . .GSB
gypsum wallboard . . . . . .GWB

**H**

hardboard . . . . . . . . . . . .HBD
hardware . . . . . . . . . . . . .HDW
header . . . . . . . . . . . . . . .HDR
heat . . . . . . . . . . . . . . . . . .HT
heated . . . . . . . . . . . . . . .HTD
heater . . . . . . . . . . . . . . .HTR
heating . . . . . . . . . . . . . . .HTG
heating, ventilating, and
    air conditioning . . . . .HVAC
height . . . . . . . . . . . . . . .HGT
hexagon . . . . . . . . . . . . .HEX.
high point . . . . . . . . . . . .HPT
highway . . . . . . . . . . . . .HWY
hinge . . . . . . . . . . . . . . .HNG
hollow-core . . . . . . . . . . . .HC
hollow metal door . . . . .HMD
honeycomb . . . . . . . .HNYCMB
horizontal . . . . .HOR or HORZ
horsepower . . . . . . . . . . . .HP
hose bibb . . . . . . . . . . . . .HB
hot air . . . . . . . . . . . . . . . .HA
hot water . . . . . . . . . . . . .HW
hot water heater . . . . . .HWH
humidity . . . . . . . . . . . . .HMD

**I**

illuminate . . . . . . . . . . .ILLUM
incandescent . . . . . . .INCAND
inch . . . . . . . . . . . . . . . . . .IN.
inch per second . . . . . . . .IPS
inside diameter . . . . . . . . . .ID
install . . . . . . . . . . . . . . .INSTL
insulation . . . . .INS OR INSUL
interior . . . . . . . . . . . . . . .INT
iron . . . . . . . . . . . . . . . . . . . .I

**J**

jamb . . . . . . . . . . . .JB or JMB
joint . . . . . . . . . . . . . . . . . .JT
joist . . . . . . . . . . . . . . . . . . . .J

**K**

kiln-dried . . . . . . . . . . . . . .KD
kitchen . . . . . . . .K, KT, or KIT.

**L**

laminate . . . . . . . . . . . . .LAM
landing . . . . . . . . . . . . . .LDG
lateral . . . . . . . . . . . . . . .LATL

lath . . . . . . . . . . . . . . . . .LTH
laundry . . . . . . . . . . . . . .LAU
laundry tray . . . . . . . . . . . .LT
lavatory . . . . . . . . . . . . .LAV
leader . . . . . . . . . . . . . . . . . .L
left hand . . . . . . . . . . . . . .LH
length . . . . . . . .L, LG, or LGTH
level . . . . . . . . . . . . . . . .LVL
library . . . . . . . . . . . . . . .LIB
living room . . . . . . . . . . . . .LR
light . . . . . . . . . . . . . . . . . .LT
light switch . . . . . . . . .LT SW
limestone . . . . . . . .LMS or LS
linen closet . . . . . . . . . . .L CL
lining . . . . . . . . . . . . . . . .LN
linoleum . . . . . . . . . . . .LINO
linoleum floor . . . . . . . . . .LF
          or LINO FLR
lintel . . . . . . . . . . . . . . .LNTL
living room . . . . . . . . . . . . .LR
local . . . . . . . . . . . . . . . .LCL
louver . . . . . . . . . .LVR or LV
low point . . . . . . . . . . . . . .LP
lumber . . . . . . . . . . . . . .LBR

**M**

main . . . . . . . . . . . . . . . .MN
makeup . . . . . . . . . . . .MKUP
manufactured . . . . . . . . .MFD
marble . . . . . . . . .MRB or MR
masonry . . . . . . . . . .MSNRY
masonry opening . . . . . . . .MO
material . . . . . . .MTL or MATL
maximum . . . . . . . . . . .MAX
median . . . . . . . . . . . . .MDN
medicine cabinet . . . . . . .MC
medium . . . . . . . . . . . .MDM
meridian . . . . . . . . . . . . .MER
metal . . . . . . . . . . . . . . .MET.
metal anchor . . . . . . . . . .MA
metal door . . . . . . . . . .METD
metal flashing . . . . . . . .METF
metal threshold . . . . . . . . .MT
mineral . . . . . . . . . . . .MNRL
minimum . . . . . . . . . . . .MIN
mirror . . . . . . . . . . . . . . .MIR
miscellaneous . . . . . . . .MISC
miter . . . . . . . . . . . . . . . .MIT
mixture . . . . . . . . . . . . .MIX.
modular . . . . . . . . . . . . .MOD
molding . . . . . . .MLD or MLDG
mortar . . . . . . . . . . . . . .MOR

**N**

National Electrical
    Code® . . . . . . . . . . . . . .NEC®
National Electrical
    Safety Code . . . . . . .NESC
natural grade . . . . . . . . . .NG
negative . . . . . . . .(−) or NEG
noncombustible . . . . .NCOMBL
north . . . . . . . . . . . . . . . . . .N
nosing . . . . . . . . . . . . . .NOS
not to scale . . . . . . . . . . .NTS

**O**

obscure glass . . . . . .OBSC GL
octagon . . . . . . . . . . . . . .OCT
on center . . . . . . . . . . . . . .OC
one-pole . . . . . . . . . . . . . . .SP
opening . . . . . .OPG or OPNG
open web joist . . . .OJ, OW J,
          or OW JOIST
opposite . . . . . . . . . . . . .OPP
optional . . . . . . . . . . . . .OPT
ordinance . . . . . . . . . . . .ORD
outlet . . . . . . . . . . . . . . .OUT.
outside diameter . . . . . . . .OD
out-to-out . . . . . . . . .O TO O
overall . . . . . . . . . . . . . . .OA
overcurrent . . . . . . . . . . . .OC
overcurrent relay . . . . . .OCR
overhead . . . . . . .OH or OVHD

**P**

paint . . . . . . . . . . . . . . . .PNT
panel . . . . . . . . . . . . . . .PNL
pantry . . . . . . . . . . . . . .PAN.
parallel . . . . . . . . . . . . . .PRL
partition . . . . . . . . . . . . .PTN
passage . . . . . . . . . . . .PASS.
penny (nails, etc.) . . . . . . . .d
perimeter . . . . . . . . . . .PERIM
perpendicular . . . . . . . . .PERP
per square inch . . . . . . . . .PSI
phase . . . . . . . . . . . . . . . .PH
piping . . . . . . . . . . . . . . . .PP
plaster . . . . . . . . . .PLAS or PL
plastered open . . . . . . . . .PO
plate . . . . . . . . . . . . . . . . .PL
plate glass . . . . .PG, PL GL,
          or PLGL
platform . . . . . . . . . . . .PLAT
plumbing . . . . . . . . . . . .PLBG
plywood . . . . . . . . . .PLYWD
point . . . . . . . . . . . . . . . . .PT
point of beginning . . . . . .POB
polyvinyl chloride . . . . . . .PVC
porch . . . . . . . . . . . . . . . . . .P
pound . . . . . . . . . . . . . . . .LB
power . . . . . . . . . . . . . . .PWR
power supply . . . . . .PWR SPLY
precast . . . . . . . . . . . . .PRCST
prefabricated . . . . . . . . . .PFB
          or PREFAB
prefinished . . . . . . . . . . . .PFN
property . . . . . . . . . . . .PROP.
property line . . . . . . . . . . . .PL
pull switch . . . . . . . . . . . . .PS
pump . . . . . . . . . . . . . . .PMP

**Q**

quadrant . . . . . . . . . . .QDRNT
quarry tile . . . . . . . . . . . . .QT
quarry tile base . . . . . . . .QTB
quarry tile floor . . . . . . . .QTF
quarter . . . . . . . . . . . . . .QTR
quarter-round . . . . . . . .¼RD

# ABBREVIATIONS

**R**

radiator ........RAD or RDTR
raised ...................RSD
random .................RDM
range ......................R
receptacle .............RCPT
recessed ...............REC
rectangle ..............RECT
redwood ................RWD
reference ...............REF
reference line .........REFL
reflected .............REFLD
refrigerator .....REF or REFR
register .....REG or RGTR
reinforce or
  reinforcing ....RE or REINF
reinforced concrete ......RC
reinforcing steel ........RST
reinforcing steel
  bar .................REBAR
required ..............REQD
retaining .............RETG
revision ...............REV
revolution per minute ...RPM
revolution per second ....RPS
right hand .............RH
riser ......................R
road ....................RD
roof .....................RF
roof drain ..............RD
roofing ................RFG
room .............RM or R
rough ..................RGH
rough opening ..........RO
            or RGH OPNG
rough-sawn .............RS
round ...................RND
rubber .................RBR
rubber tile .....RBT or R TILE
rustproof ............RSTPF

**S**

safety ..................SAF
sanitary ..................S
S-beam ....................S
scale .....................SC
schedule ......SCH or SCHED
screen....SCN, SCR, or SCRN
screen door ............SCD
screw ..................SCR
scuttle....................S
section ........SEC or SECT.
select ..................SEL

self-cleaning ........SLFCLN
self-closing ..........SELF CL
service ........SERV or SVCE
sewer ..................SEW.
sheathing ...SHTH or SHTHG
sheet ............SHT or SH
sheeting ...............SH
sheet metal .............SM
shelf and rod ........SH&RD
shelving ........SH or SHELV
shingle ................SHGL
shower ..................SH
shower and toilet .....SH & T
shower drain ............SD
shutter ...............SHTR
sidelight ..............SI LT
sill cock .................SC
single-phase ...........1PH
single-pole ..............SP
single-pole double-
  throw ..............SPDT
single-pole double-throw
  switch ..........SPDT SW
single-pole single-
  throw ..............SPST
single-pole single-
  throw switch ....SPST SW
single-pole switch .....SP SW
single-strength glass ....SSG
single-throw..............ST
sink ...............SK or S
skylight ................SLT
sliding door ....SLD or SL DR
slope ...................SLP
soffit ...................SF
soil pipe..................SP
soil stack ...............SSK
solid core ...............SC
soundproof .........SNDPRF
south ....................S
specific...................SP
specification ..........SPEC
splash block ............SB
square ...................SQ
square feet .........SQ FT
square inch .........SQ IN.
square yard ........SQ YD
stack ...................STK
stained ................STN
stainless steel ........SST
stairs ....................ST
stairway ..............STWY
standard ..............STD
steel ............ST or STL

steel sash.................SS
stone ...................STN
storage ........STO or STG
street ...........ST or STR
structural .............STRL
Structural Clay Products
  Research Foundation ...SCR
structural clay tile .....SCT
structural glass ........SG
supply ................SPLY
survey .................SURV
suspended ............SUSP
switch ..........SW or S

**T**

telephone ...............TEL
television ...............TV
temperature ..........TEMP
tempered plate
  glass .........TEM PL GL
terra cotta ..............TC
terazzo .........TZ or TER
thermostat .........THERMO
thick ..................THK
threshold ..............TH
tile base ...............TB
tile drain ...............TD
tile floor ...............TF
timber ................TMBR
toilet .....................T
tongue-and-groove ....T & G
township ..................T
tread ............TR or T
typical .................TYP

**U**

underground .........UGND
unexcavated ........UNEXC
unfinished ....UNFIN or UNF
unit heater .............UH
unless otherwise
  specified ...........UOS
untreated ............UTRTD
utility ..........U or UTIL
utility room .....UR or U RM

**V**

vacuum ................VAC
valley ..................VAL
valve.....................V
variance ...............VAR
vent ......................V
vent hole ...............VH
ventilate ............VENT.

ventilating equipment ....VE
vent pipe ................VP
vent stack ...............VS
vertical ................VERT
vestibule ..............VEST.
vinyl tile .......VT or V TILE
vitreous tile........VIT TILR
void .....................VD
volt .......................V
voltage ...................V
voltage drop ............VD
volt-amp ................VA
volume ................VOL

**W**

wainscot ......WSCT, WAIN.,
              or WA
walk-in closet...........WIC
wall ......................W
wallboard ..............WLB
wall receptacle .........WR
warm air ...............WA
washing machine.......WM
water ..........WTR or W
water closet ...........WC
water heater ...........WH
water line ..............WL
water meter ...........WM
waterproof.........WTRPRF
water-resistant .........WR
watt ......................W
weatherproof .......WTHPRF
              or WP
weather-resistant ........WR
weather stripping ........WS
weep hole ...............WH
welded wire fabric .....WWF
west .....................W
white pine ..............WP
wide .....................W
wide flange ........W or WF
window ...............WDO
wood ..................WD
wood frame ...........WF
wrought iron...........WI

**Y**

yard ...................YD
yellow pine ............YP

**Z**

zone .....................Z

## A

**A-frame.** Building with a gable roof extending to the foundation.

**Abbreviations.** Key letters of words denoting the complete word. Most are standardized by standards organizations.

**Accordion door.** Door made of wood slats or fabric arranged to fold back and forth in the plane of the door frame.

**Acoustical tile.** Ceiling tile with small holes that trap sound to reduce the reflection of sound.

**Acute triangle.** A triangle with no 90° or greater angles.

**Actual size.** True size of lumber in contrast to nominal size. The actual size of a 2 × 4 is $1\frac{1}{2}'' \times 3\frac{1}{2}''$.

**Additive.** Chemical added to concrete to alter its properties. Accelerators, retardants, and other entraining agents are examples of additives.

**Aggregate.** Gravel, broken stone, or other hard inert material used in concrete.

**Air duct.** Pipe, usually rectangular or round and made of sheet metal, used to conduct hot or cold air in heating or cooling systems.

**Amp.** Measurement of electric current. Designated as *I* in formulas. Abbreviated as *A*. For example, 20 A CB.

**Anchor bolt.** Metal bolt used to secure a wood sill to a masonry or concrete foundation wall.

**Angle.** The number of degrees between two intersecting lines of a flat plane. For example, 45°.

**Angle iron.** Piece of structural steel formed with cross-sectional shape of a right angle.

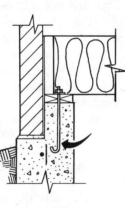

anchor bolt

**ANSI.** American National Standards Institute.

**Apron.** **1.** In concrete, the flat concrete slab in front of a garage. **2.** In carpentry, an inside trim member on a window.

**Architect.** Person qualified and licensed to design and oversee construction of a building.

**Asphalt shingle.** Shingle made of felt saturated with asphalt and a surface covered with mineral granules. Used as a finish roofing material.

**Attic.** Space between the ceiling and roof of a building.

**Axonometric.** Pictorial drawing in which the axes vary depending upon the type. Three types are isometric (most common), dimetric, and trimetric.

## B

**Backfill.** Coarse earth, or other material, used to build up the ground level around foundation walls or in low areas.

**Backsplash.** The rear vertical piece of a countertop that is placed along a wall.

**Balloon framing.** Framing method in which studs extend from the sill plate to the roof. Second floor joists, which are spiked into the studs, receive their main support from a ribbon notched into and nailed to the studs.

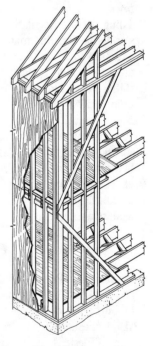

balloon framing

**Baseboard.** Molding placed at the base of a wall and fitted to the floor. Also called *base molding*.

**Base cabinet.** Kitchen cabinet placed against a wall and resting on the floor. Standard size of base cabinets is 24″ deep and 36″ high. The length varies.

**Baselines.** East-West lines in the gridwork of lines that crisscross the USA and form townships.

**Base shoe.** Molding placed against a baseboard at the floor.

**Batt insulation.** Blanket insulation placed between studs. Common widths fit between studs on 16″ or 24″ centers.

**Batten.** A narrow strip of wood used to cover the joint between two pieces of vertical siding.

**Bench mark.** In surveying, a mark on some object firmly fixed in the ground from which distances and elevations are measured. It is usually a mark established by the local government as a local point of reference.

**Bevel.** An angled cut from surface to surface of a board.

**Beveled siding.** Siding that is tapered from one edge of the board to the other.

**Bid.** Offer to perform work for a specified price.

**Board foot.** Unit of measure for lumber based on the volume of a piece 12″ square and 1″ thick. A board foot contains 144 cu in. (12″ × 12″ × 1″ = 144 cu in.).

**Branch circuit.** In electrical work, the circuit conductors between the final overcurrent protection device and outlet.

**Brick bond.** Pattern formed by exposed faces of brick.

**Brick veneer construction.** Frame construction with a brick wall or other masonry units.

**Bridging.** Bracing between joists or studs that adds stiffness to the floors and walls.

**Buck.** Frame placed inside a concrete form to provide an opening for a door or window.

**Building brick.** Brick made from common clay without a special surface treatment. Also called *common brick*.

**Building code.** Regulations that establish required standards for the materials and methods of construction in a city, county, or state. Building codes are enforceable by law.

**Building lines.** Lines laid out to establish faces of exterior walls.

**Building permit.** Legally required authorization for construction work.

**Built-up girder.** Girder made of laminated wooden boards designed to carry heavy loads.

**Butt.** A hinge for a door.

**Butt joint.** Joint in which one piece butts squarely against another.

# C

**Cabinetmaker.** Person who works in a cabinet shop and is skilled in the layout, construction, and installation of wood cabinets.

**CAD (computer-aided drafting).** Graphic representation of designs using computers. CAD also stands for *computer-aided design*.

**Cased opening.** Finished interior opening without a door.

**Casing.** A wood trim member covering the space between plaster or drywall and the jamb at windows or doors.

**Caulk.** Nonhardening paste used to fill cracks and crevices.

**Ceiling grid.** Light metal framework supporting tiles of a suspended ceiling.

**Ceiling tile.** Rectangular or square fibrous pieces used to finish off ceilings.

**Ceiling heating.** System in which heat comes from a single source and is distributed by ducts or pipes in the ceiling to all parts of a building.

**Central processing unit (CPU).** The control center of a computer. For CAD, the CPU receives information from an input device, manages the information, and produces an output image.

**Ceramic tile.** Thin flat piece of fired clay attached to walls or floors.

**Chamfer.** An angled cut from a surface to an adjacent edge of a board.

**Chord.** **1.** In framing, the horizontal member of a truss, commonly the bottom member. **2.** In drafting, a line from circumference to circumference of a circle that does not pass through the centerpoint.

**Circle.** A plane figure generated about a centerpoint. All circles contain 360°.

**Circuit.** Complete electrical path. Dashed lines on plans show circuit wiring. Solid lines with arrowheads indicate home runs.

**Circumference.** The outside boundary of a circle.

**Closed stringer.** A stair stringer with routed grooves for the ends of treads and risers which are concealed in the finished stair.

circuit

**CMU (concrete masonry unit).** Building material such as cinder block or concrete block.

**Common nail.** Flat-head nail used most often in rough work.

**Common rafter.** Roof member extending from the top wall plate to the ridge of a gable roof.

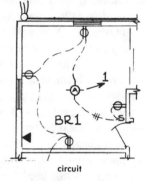
common rafter

**Compass direction.** Direction based on the compass points North, South, East, or West. Commonly used on plot plans to orient the house on the lot. Also used to designate the various exterior elevation views.

**Concrete.** A mixture of cement, sand, and gravel with water in varying amounts according to the use of the finished product.

**Concrete block.** Precast block, solid or hollow, used in the construction of walls. Also known as *CMU*.

**Concrete mix.** The proportion of cement, sand, and gravel in a mixture of concrete.

**Conductor.** **1.** In electrical work, any material that conducts electricity. NEC® Table 310-16 gives ampacities of copper, aluminum, and copper-clad aluminum conductors. **2.** In plumbing, a pipe that carries water to the ground or storm sewer. Also known as a *leader* or *downspout*.

**Conduit.** Metal tubing used to carry electrical conductors.

**Contour lines.** Lines on a plot plan drawn to pass through points having the same elevation. Dashed lines represent natural grade. Solid lines represent finish grade.

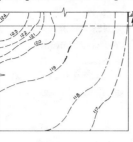

contour lines

**Convenience outlet.** Electrical outlet in the wall or floor that can be used for lamps and appliances.

**Convention.** Simplified way of representing a building component on prints.

**Cord- and plug-connected.** Wiring method in which equipment is connected with a cord and plug to a receptacle. For example, a range may be cord- and plug-connected.

**Crawl space.** Space between the ground and bottom of the joists in a house that does not have a basement.

**Cross bridging.** Wood or metal bridging placed diagonally between joists.

**Cutting plane.** Line (identified by letters) that cuts through a part of a structure on a drawing. It refers to a separate elevation, sectional view, or detail drawing given for that area.

## D

**Dead load.** Weight of the permanent structure of a building, including all materials that make up the unit.

**Degree.** In plane measurement, 1/360th of a circle. Used to designate angles between planes or lines. Designated °. For example, 45°.

**Detail.** Scaled plan, elevation, or sectional view drawn to a larger scale to show special features.

**Diameter.** Distance from circumference to circumference of a circle through the centerpoint.

**Diazo process.** Process of producing prints that have blue or black lines on a white background.

**Door hand.** Direction in which a door swings. For example, a right-hand door is hinged on the right and swings away from a person standing outside the door.

**Dormer.** Projection from a sloping roof that provides additional interior area. Three common types are gable-end, hipped-end, and shed.

**Drywall.** A system of interior wall finish using sheets of gypsum board with taped joints.

**Duct.** 1. In HVAC, a large round or rectangu-

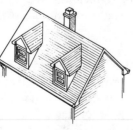

dormer

lar pipe used for carrying air. 2. In electrical work, a rectangular-shaped trough that serves as a wireway.

## E

**Easement.** Privately owned land used for public utilities.

**Elevation.** 1. In printreading, the orthographic view of the exterior or interior walls. 2. In measurement, the height of a point above sea level or some datum point. For example, 328'-0".

**Ellipse.** A plane figure generated by the sum of the distances from two fixed points.

**Equilateral triangle.** A triangle with three equal sides and three equal angles.

## F

**Fascia.** Flat outside horizontal member of a cornice placed in a vertical position.

**Fill insulation.** Granulated mineral wool or pellets made from substances such as glass, slag, rock, and expanded mica.

**Finish grade.** Various levels of the lot surface after final grading work has been completed. Represented on plot plans by solid lines.

**Finish hardware.** Hardware that is visible, such as hinges, locks, catches, door stops, door closers, and coat hooks.

**Fire cut.** Angled cut in the end of a joist that allows a burnt joist to fall out of a brick wall without disturbing the wall.

**Flashing.** Sheet metal used in roof and wall construction to make them waterproof.

**Flexible metal conduit.** Electrical conduit consisting of a spiral-wound steel strip.

**Floor joist.** Common joist supporting the rough floor material.

**Floor plan.** Scaled view looking directly down on the dwelling. The cutting plane is taken 5'-0" above the finished floor.

**Flush door.** Door with a flat surface on both sides.

**Foundation.** The part of a building resting on and extending into the ground that provides support for structural loads above.

**Foundation footing.** 1. The part of a foundation resting on bearing soil and supporting the foundation wall. 2. The base for a column.

**Foundation plan.** Drawing in a set of prints giving a plan view of the foundation of a building.

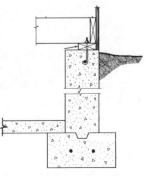

foundation

**Furnace.** A device fueled by fuel oil, natural gas, or electricity that heats air in a forced warm air heating system.

**Furring.** Strips fastened to a wall, floor, or ceiling to which covering material is attached.

# G

**Gable roof.** Roof that slopes in two directions from the ridge. Most common type of roof used in residential construction.

**Gambrel roof.** Roof with a double slope in two directions from the ridge.

**Gauge.** **1.** In metalwork, a uniform standard for wire diameters and thicknesses of metal sheets. For example, 16 ga. **2.** In concrete work, proportion of the various materials in mortar or plaster mix.

**General contractor.** Licensed individual or firm in charge of a construction project.

**GFCI (ground-fault circuit interrupter).** A device used to detect leakage of current and trip the circuit open at 5 milliamps or less.

**Grade.** The level of the ground around a building.

# H

**Hardware.** **1.** Physical components of a computer system. Common CAD hardware items are the keyboard, graphics tablet, display monitor, and plotter. **2.** Common term for hinges, locks, etc. of doors and windows.

**Header.** **1.** Joist placed at the ends of an opening in the floor used to support side members. **2.** The top rough framing member over a window or door opening.

**Heat pump.** Combined heating-cooling system that extracts heat from outdoor air and transfers it indoors in winter. In summer, it reverses the cycle, absorbing heat indoors and sending it to the outside.

**Hexagon.** A plane figure with six sides. In a regular hexagon, each angle is 60°.

**Hip roof.** Roof that slopes in four directions from the ridge.

**Hollow-core door.** Door made of wooden strips glued together on edge like an eggcrate and covered with veneer plies.

**Horizontal.** Level or parallel with the horizon.

**HP (horsepower).** Unit of power equal to 746 watts. Designated HP. For example, 3.5 HP.

**HVAC.** Heating, ventilating, and air conditioning.

# I

**Input device.** Hardware that enters information into a CAD system. Common input devices include keyboard, stylus, and mouse.

**Interior elevation.** Scaled view that shows the shape and size of interior walls and partitions of a house.

**Irregular polygon.** Plane figures with sides of varying lengths.

**Isosceles triangle.** A triangle with two equal sides and two equal angles.

**Isometric.** Type of pictorial drawing in which horizontal axes are drawn at a 30° angle from the horizontal. The isometric axis is drawn on a 120° angle.

# J

**Jamb.** Main member of a window or door frame, forming the sides and top.

**Joist.** Framing member that directly supports the floor.

# K

**Keyway.** Groove in one lift of concrete that is filled with concrete of the next lift.

# L

**Lateral.** In electrical work, underground conductors that connect the power company's transformer to the service equipment.

**Level.** **1.** Horizontal. **2.** A builder's instrument that revolves only in a horizontal plane, used to transfer points in laying out foundations. **3.** Tool used to level building parts during construction. **4.** To adjust into a horizontal position.

**Live load.** All loads that are not part of the structure, such as people, furnishings, snow, and wind.

**Location dimension.** Dimension that locates a feature in relation to another feature.

**Lockset.** Keyed door hardware with rectangular, beveled, or rectangular and beveled bolts.

**Longitudinal section.** Sectional view created by passing a cutting plane through the long dimension of a house.

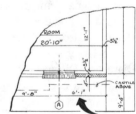

location dimension

**Lookout.** Horizontal wood structural member projecting beyond the face of the building.

**Loose-fill insulation.** Insulation poured directly from a bag or blown into place with a pressurized hose.

# M

**Main.** **1.** In electrical work, the primary overcurrent protection device that protects busbars in a panelboard. **2.** In plumbing, the major supply pipe for gases or liquids.

**Mansard roof.** Roof with a double slope in four directions from the ridge.

**Masonry.** Construction using molded or shaped construction material such as concrete blocks, bricks, stones, and tiles.

**Masonry veneer.** Exterior finish cover of brick or stone, usually applied over a wood stud wall.

**Meridians.** North-South lines in the gridwork of lines that crisscross the USA and form townships.

**Mil.** One thousandth of an inch (.001″).

**Minute.** As related to plane measurement, 1/60th of a degree. Designated ′. For example, 30°-58′.

**Modular brick.** Brick classified by its actual size and designed so that every third horizontal joint falls on a multiple of 4″.

# N

**Natural grade.** Levels of the lot surface before finish grading.

**Nominal size.** Descriptive size, not actual measured size. For example, 2″ × 4″ is the nominal size of a piece of wood actually measuring 1¹/₂″ × 3¹/₂″.

**Non-load-bearing wall.** Wall that supports its weight only.

**Nonmetallic-sheathed cable.** Electrical cable containing two to four conductors with a nonmetallic jacket protecting the conductors.

# O

**Oblique.** A type of axonometric drawing with one surface shown as a true (normal) view and having receding lines of 30° or 45°. Obliques may be cabinet or cavalier drawings.

**Orthographic.** Method of projecting planes at right angles.

**Overcurrent protection device.** Either fuses or a circuit breaker (CB) used to protect circuit conductors from overload.

# P

**Panel door.** Door with solid strips joined together to hold panels.

**Panelboard.** Metal enclosure that houses overcurrent protection devices.

overcurrent protection device

**Panned ceiling.** Ceiling with two levels connected by sloped surfaces.

**Penny.** Measurement of nail length. Abbreviated as *d*. For example, an 8d nail is 2¹/₂″ long.

**Pentagon.** A plane figure with five sides. In a regular pentagon, each angle is 72°.

**Perspective.** Type of pictorial drawing in which receding lines converge.

**Pictorial drawing.** A drawing that shows three surfaces of an object.

**Pitch (Slope).** The slope of a roof expressed as a ratio of rise to run. For example, an 8 in 12 roof rises 8″ vertically per 1′-0″ of horizontal measurement.

**Plan view.** 1. A view looking down. 2. In orthographic projection, a top view. 3. In architecture, a floor plan, roof plan, or plan view of a cabinet.

**Plane figure.** A flat figure.

**Plank-and-beam framing.** System of wood-frame construction in which planks and beams provide structural support.

**Platform framing.** Wood-frame construction in which studs are one story high. A platform is built on plates over the studs and acts as a base for the next floor. Also called *western framing.*

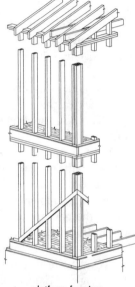

platform framing

**Plot plan.** Scaled view that shows the shape and size of the building lot and the location, shape, and overall size of the house on the building lot.

**Plumb.** In a vertical position.

**Plywood.** Product made of veneer sheets glued at right angles to one another and pressed together under high heat and pressure. Plywood always has an odd number of plies.

**Point of beginning.** The point on a lot from which horizontal and vertical measurements are made.

**Polygon.** A plane figure generated about a centerpoint. A regular polygon has sides of equal lengths.

**Print.** Reproduction of a working drawing.

**Property lines.** Recorded legal boundaries of a piece of property.

# Q

**Quadrant.** One-fourth of a circle.

**Quadrilateral.** Plane figure with four sides.

# R

**Raceway.** Enclosed channel of metal or insulating material designed to protect electrical conductors.

**Radius.** One-half the diameter of a circle.

**Rafter.** A sloping roof member that supports the roof sheathing.

**Rebar.** Steel bar used for reinforcing concrete structural members.

**Receptacle.** Electrical contact device installed at the outlet for the connection of a single attachment plug.

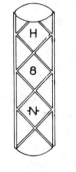

rebar

**Rectangle.** Quadrilateral with opposite sides that are equal and four 90° angles.

**Register.** A grill through which heated or cooled air flows into or out of a room.

**Rhomboid.** Quadrilateral with opposite sides equal and no 90° angles.

**Rhombus.** Quadrilateral with four equal sides and no 90° angles.

**Retaining wall.** Masonry or wood wall constructed to hold back earth.

**Ridge.** Top horizontal member of a roof framed with rafters.

**Rise. 1.** In carpentry, the vertical measurement from the support to the ridge of the roof. For example, a rafter may have a 3″ rise per foot of run. **2.** In stairs, the vertical measurement from the top of a tread to the top of the next higher tread.

**Riser. 1.** In carpentry, the vertical part of a stairstep. **2.** In plumbing, a vertical water supply line. **3.** In HVAC, a vertical heating supply duct.

**Right triangle.** Triangle with one 90° angle.

**Rough grade.** Various levels of the lot before grading has been completed.

**R-value.** In insulation, the resistance to heat flow. The higher the R-value is, the better the insulating qualities of the material are. For example, R-19.

# S

**Section.** A subdivision of a township. A section is one mile long on each side.

**Sector.** Pie-shaped portion of a circle.

**Scuttle.** An opening in a ceiling providing access to an attic.

**Sectional view.** Scaled view created by passing a cutting plane through a portion of a building.

**Setback.** Required distance a building must be placed from a given boundary as established by ordinance.

**Sheathing.** First layer of exterior wallcovering applied over the framing. Types of sheathing include fiberboard, gypsum board, and plywood.

**Sheetrock.** Trade name for gypsum wallboard.

**Size dimension.** Dimension that gives the size of an area or feature.

**Slab-on-grade.** Ground-supported concrete foundation system consisting of foundation walls and concrete slab.

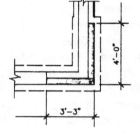

size dimension

**Slope (Pitch).** Relationship of roof rise to run.

**Soffit.** A lower horizontal surface such as the underface of eaves, cornice, or beam.

**Software.** Graphics, word-processing, or data processing programs for computers.

slope

**Solar energy.** Energy from the sun's rays.

**Solid bridging.** Bridging made of solid blocks of wood placed between joists.

**Solid-core door.** Flush door made of solid wood blocks glued together and covered with veneer plies. Can be used as interior or exterior door. Building codes require that exterior doors be solid-core.

**Span.** Total clear distance between supports.

solid bridging

**Specifications.** Written supplements to working drawings that give additional building information.

**Square. 1.** A quadrilateral with four equal sides and four 90° angles. **2.** In layout, a 90° angle. **3.** In roofing, the amount of roofing that will cover 100 sq ft when laid.

**Stair flight.** Section of stairs going from one floor or landing to another.

**Stair landing.** Platform between one flight of stairs and another.

**Stair pitch.** The slope of a set of stairs.

**Stair ratio.** Ratio between the unit tread and unit rise of a set of stairs expressed as a formula ($T + R = 17″$ to $18″$).

**Staircase.** Entire assembly of stairs, landings, railings, and balusters.

**Stairwell.** Opening in the floor provided for stairs.

**Stepped foundation.** Foundation with a change in elevation to provide for a change in grade or special building condition.

**Stud.** Vertical structural upright used to form the walls and partitions in a frame building.

**Survey.** A plan of a lot prepared by a licensed surveyor. Shows lot dimensions, angles at corners, elevations, and other data.

**Survey plat.** A plan showing land divisions of townships, streets, and lots.

**Swale.** The slope on a lot that ensures water drainage away from the building.

**Symbols.** Graphic representations of building materials.

# T

**Take-off.** Estimate of the amount of material required for a job.

**Title block.** Identification section on the right side or bottom of each sheet in a set of plans. Title blocks give the number of the sheet, total number of sheets, names of owner, architect, drafter, and checker, and the scale and lot number.

**Township.** An area six miles long on each side. Townships are divided into sections and quarter sections.

**Topsoil.** Uppermost layer of soil that is capable of supporting vegetation.

**Transverse section.** Sectional view created by passing a cutting plane through the short dimension of a house.

**Trapezoid.** Quadrilateral with two sides parallel and no 90° angles.

**Trapezium.** Quadrilateral with no sides parallel.

**Tread.** The horizontal portion of a stairstep.

**Triangle.** Plane figure with three sides. All triangles contain 180°.

# U

**Underlayment.** Floor covering of plywood or fiberboard used to provide a level surface for carpet or other resilient flooring.

**Unit rise.** **1.** In roofing, the number of inches a common rafter rises vertically for each foot of run. **2.** In stairs, the riser height. Calculated by dividing the total rise by the number of risers.

**Unit run.** **1.** In roofs, the unit of total run based on 12″ (17″ for hip roofs). **2.** In stairs, the width of the tread calculated by dividing the total run by the number of treads.

**Utilities.** Electric, gas, water, and sewage services provided to the public.

# V

**Valley.** Angle formed by two inclined sides of a roof. A valley rafter is the rafter supporting the valley.

**Vanity.** Base cabinet with a lavatory. Standard size of vanities is 20″ deep and 30″ high.

**Vapor barrier.** Watertight material used to prevent passage of moisture or water vapor into walls or slabs.

**Veneer.** **1.** A thin layer of wood. **2.** In masonry, a facing of brick, stone, or other units placed over wood framing.

**Vent stack.** Vertical pipe connected to the plumbing vent pipes and soil waste stacks to remove gases and relieve pressure in the system.

**Vertical.** In a plumb or upright position.

**Volt.** Pressure that forces the flow of electrons in an electrical circuit. Designated $E$ in power formula. Abbreviated as $V$.

**Volt-amp.** Electrical power equal to *volts* × *amps*. Abbreviated as *VA*.

# W

**Watt.** Measurement of electrical power. Designated $P$ in power formula. Beginning with 1984 NEC®, termed *VA* (volt-amps) for calculations.

**Welded wire fabric.** A mesh made of heavy wire in a rectangular or square pattern welded at intersections of the wire. Used to reinforce concrete.

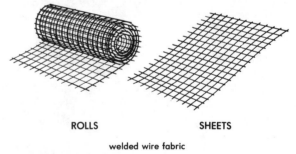

ROLLS        SHEETS

welded wire fabric

**Working drawings.** Sets of plans that contain the complete graphic information necessary to complete a job.

**Wythe.** Single, continuous masonry wall one unit thick.

## A

Abbreviations, 46, *49*, 85, 204–206
  on floor plans, 63
  on plot plans, 136, *137*
AC armored cable, 153–154
Acute triangle, 20, *21*
Air conditioning, 164–166, *166*
  floor plans, 165
  package system, 164–165, *166*
  specifications, 165
  split system, 164, *166*
Air registers, 162, *163*
American National Standards
  Institute, 28
ANSI. *See* American National
  Standards Institute
Arc, 20, *21*
Architect's scale, 9, *9*, 26–27, *28*
Architectural symbols, *192–193, 194*
Awning windows, 120, *123*

## B

Balloon framing, 102, *104*
Baselines, 134
Basement
  unfinished, 71
  walkout, 70
Bath
  elevation detail, *118*, 119, *120*
  plan view, 118, 119, *120*
Bathrooms, 72
  partitions in, 30
Batterboards, 145, *146*
Bench mark, 135
Blowers, 162
Blueprints, 4, *5*, 6. *See also* Prints
Bottom plate, 146
Branch circuit, 152, *153*
Branch circuit conductors, 152
Breakfast area, 72
Brick, 150–151, *201*
  symbol for, *49, 85*
Brick bond, 151
Brick veneer construction, 100, 105,
  *105*
Brick veneer wall, 30, *30, 47*
  openings in, *48*
Brick wall, *47*
  openings in, *48*

Bridging, 147–148, *148*
Building codes, 133–134, *134*
Building design, 85–89
Building permits, 133, *134*
Built-in-place forms, 146, *146*
Built-up girder, 68
Butted joists, 147

## C

Cables, 153–155
CABO. *See* Council of American
  Building Officials
CAD. *See* Computer-aided design
CADD. *See* Computer-aided design
Cape Cod house, 69
Carpentry, 145–150
CCA. *See* Chromated copper arsenate
Chimney, 68
Chord, 20, *21*
Chromated copper arsenate, 90
Circles, 20, *21*
Circumference, 20, *21*
Civil engineer's scale, 9, *9*, 134
Common brick, *85*
Compasses, 8, *8*
Computer-aided design, 9–12, *10, 11,*
  *12*
  drawings, of floor plans, 67
  input systems, 10–11, *10*
  output systems, 11, *11*
  plans, 11–12, *12*
Computer software programs, 20
Concrete block, *47*, 150–151, *201*
  dimensioning, 30
  symbol for, *49, 85*
Concrete foundation work, 145–147
Concrete reinforcement, *202*
Conductors, 151
  branch circuit, 152
  feeder, 152
  service-entrance, 151–152
Contour lines, 2, 4, 135, *136*
  closely spaced, 137
  dashed, 4
  numbers on, 137
  solid, 2, 4
Cornice detail, 120, 122, *123*
Council of American Building
  Officials, 133
Cross bridging, 147, 148

Cut stone, symbol for, *85*
Cutting plane line, direction arrow
  and, 2
Cutting planes, 1, 2, 65, *65*, 67, 69, 117

## D

Denominator, 134
Details, structural, 119–120, *121*
Detail views, 1, *3*, 99, 115–124, *116*
  defined, 115
  dimensions on, 117
  elevation, 115
  of exterior trim, 115
  full-size, 117
  of interior trim, 115
  of kitchen, 117, *119*
  reading, 122–124
  structural, 119–120
  typical, 99–100
Diameter, 20, *21*
Diazo prints, 4, *5*, 6
  dry method of producing, 6
  wet method of producing, 6
Dimensioning, 28, *29*
Dimensions
  on detail views, 117
  on floor plans, 63
  location, 117
  size, 117
Dividers, 8, *8*
Door details, 120
Door hand, 89
Doors, 86, 87, 89, *89*
  accordion, *49*
  dimensioning, 30, *30*
  flush, 87
  hollow-core, 87
  panel, 87
  sliding, *49*
  solid-core, 87
  symbols for, *49, 194*
Dormers, 69, *70*
Double-hung windows, 120, *123*
  symbol for, *49*
Drafting instruments, 8–9, *8*
Drafting methods, 6–12
  conventional, 6–9, *7*
Drainage piping, 157–158
Drawing symbols, 28, *29*
Driveways, 147

Ductwork, supply air, 162, *162*
Duplex nails, 147

**E**

Easement, 134
   utility, 136
Electrical, 151–155
Electrical metallic tubing, 152
Electrical nonmetallic tubing, 152–153
Electrical symbols, *195–196*
Electrostatic prints, 4, *5,* 6, *8*
Elevations, 134–135
Elevation views, 2, *3,* 83–90, *84, 87*
   exterior, 2
   information on, 83, 85
   interior, 2
   reading, 89–90
   of window, *118*
Entry, 71
Equilateral triangle, 20, *21*
Exterior elevations, 2
   and floor plans, 65
Exterior finish, 89
Exterior trim, 149
Exterior walls, dimensioning 28–29, *29*

**F**

Face brick, symbol for, *47, 85*
Feeder conductors, 152
Finish, exterior of house, 89
Finish grade, 135, *136*
Fireblocking, 102
Fire cut, 105
Fireplace, *47,* 68, 120, *122*
   foundation, 69
Flashing, sheet metal, *85*
Flat roof, 86, *86*
Floor joists, 147, 148
Floor plans, 1–5, *5,* 63–72, *64. See
      also* Plan view
   abbreviations on, 63
   for air conditioning, 165
   CAD drawings of, 67
   defined, 63
   dimensions on, 63
   electrical, 155, *156*
   and exterior elevations, 65
   notes on, 63–64
   of one-story house, 67, *68*
   plumbing, 158, *160*
   reading, 69–72
   simplified, 66–69
   symbols on, 63
Floors, 147–149, *148*
Flush doors, 87
Footings, 69, 145, *146*

Forced warm air heating, 161–163, *161*
Foundation work, 145–147
Frame wall, 30, *30*
   openings in, *48*
   symbols for, 47
Furnaces, 162

**G**

Gable roof, 86, *86*
Gambrel roof, 86, *86*
Garage, 72
Garage floors, 147
Geothermal heating, 164
GFI receptacles, 119
Girders, 147
   built-up, 68
Glass, symbol for, *85, 192*

**H**

Hallway, 71
Header joists, 147
Heating
   forced warm air, 161–163, *161*
   geothermal, 164
   hot water, 163–164, *163*
   radiant electric, 164, *164*
   solar, 164, *165*
Heating ducts, 65
Heating, ventilating, air conditioning.
      *See* HVAC
Hip roof, 86, *86*
Hollow-core doors, 87
Hot water heating, 163–164, *163*
HVAC, 160–166
HVAC symbols, *199*

**I**

ICBO. *See* International Congress
      of Building Officials
Insulation, *193*
Interior elevations, 2
Interior partitions
   dimensioning, 30, *30*
   openings in, *48*
Interior trim, 149
Interior wall elevations, 117, *119*
International Congress of Building
      Officials, 133
Isometric drawings, 22, *24*
Isosceles triangle, 20, *21*

**J**

Joists, 68, 147–149

**K**

Keyway, 145
Kitchen, 68, 72
   detail views, 117, *119*

**L**

Lath, *192*
Laundry, 72
Lift, 145
Light, window, 87
Lines
   alphabet of, *200*
   horizontal, 20
   slanted, 20
   vertical, 20
Living room, 72
Load-bearing partition, 68
Location dimensions, 117
Longitudinal sections, 99, 100, *101*
Lumber sizes, *203*

**M**

Mansard roof, 86, *86*
Masonry, 150–151
Masonry construction, 102, 105, *106*
Masonry walls, 30, *30,* 150, *150*
   dimensioning, 29
Mechanical engineer's scale, 9, *9*
Meridians, 134
Modular brick, 151
Moldings, 122, *191*

**N**

Nails, duplex, 147
National Electrical Code®, 151
National Electrical Safety Code, 151
Natural grade, 135, *136*
NEC®. *See* National Electrical Code®
NESC®. *See* National Electrical Safety
      Code
Nominal size, 102
Nonmetallic-sheathed cable, 154
Notations, on plans, 63–64, 117, *118*
Numerator, 134

**O**

Oblique drawings, 22, *25*
Obtuse triangle, 20, *21*
One-and-one-half story house, floor
      plans of, 69, *71*
One-story house, floor plan of, 66, *68*

Openings
  door, 86
  window, 86
Operating controls, and forced warm
    air heating, 162–163
Orthographic projections, 22, 24–26,
  *26*

**P**

Package system air conditioning,
    164–165, *166*
Panel doors, 87
Panel forms, 146–147, *146*
Partitions, 30, *30*
  load-bearing, 68
Pencils, drafting, 9, *10*
Perspective drawings, 22, *23*
Pictorial drawings, 22, *23*
Pictorial view, of window, *118*
Pipe fitting symbols, *198*
Piping, 155–158, *157*
  drainage, 157–158
  roughing-in, 158, *159*
  supply water, 155–156
  vent, 158
  waste, 156–157, *158*
Piping drawings, 158, 160, *160*
Plane figures, 20–22, *21*
Plank-and-beam framing, 102, 104,
  *104*
Plan views, *84, 87, 101*
Plaster, *192*
Platform framing, 102, *102*
Plat, survey, 133–134, *135*
Plot plans, 2, 4, *4,* 133–137
  reading, 136–137
  symbols on, *193*
Plotter, pen, 11, *11*
Plumbing, 155–160
  floor plans for, 158, *160*
  specifications for, 158, *160*
  symbols, *197*
Point of beginning, 134–135
  and elevations, 134–135
Polygons, *21,* 22
Porch, 72
Post-and-beam construction, 104
Potable water, 155–156
Power controls, and forced warm air
    heating, 162
Prints, 4–6. *See also* Blueprints
  defined, 4

**Q**

Quadrant, 20, *21*
Quadrilaterals, 20–22, *21*

**R**

Raceway systems, 152–155, *154*
Radiant electric heating, 164, *164*
Radius, 20, *21*
Rebars, 145, *193,* 202, *202*
Rectangles, 20–22, *21*
Residential construction, 100, 102–105
Rhomboid, *21,* 22
Rhombus, *21, 22*
Right triangle, 20, *21*
Rigid nonmetallic conduit, 153
Rise, 86
Riser, stair, 66
Roof framing, 149
Roofs, 86, *86,* 147–149, *150*
  shingles, *192*
  slope, 86, *87*
  trusses, 120, *122*
Rough-in sheet, for piping, 158, *159*
Run, 86

**S**

Safety controls, and forced warm air
    heating, 163
Scales, 9, *9,* 26
  architect's, 9, *9,* 26–27, *28*
  civil engineer's, 9, *9,* 134
  for detail views, 115–117, *116*
  drawing, 64–65
  for floor plans, 64–65
  mechanical engineer's, 9, *9*
Schedules, electrical, 155, *156*
Scuttle, 71
Section
  township, 134
  typical, 99–100, *100*
Sectional detail view, of window, *118*
Sectional views, 2, *3,* 99–106
  defined, 99
  reading, 105–106
Sector, 20, *21*
Semicircle, 20, *21*
Service-entrance cable, 154–155
Service-entrance conductors,
    151–152
Shed roof, 86, *86*
Sheet metal flashing, symbol for, *85*
Sheet metal work, 166, *166*
Shingles, symbol for, *85*
Siding, wood, *85*
Size dimensions, 117
Sketching, 19–26
  techniques, 19–22
Slab-on-grade foundations, 147, *147*
  and hot water heating, 163
Solar heating, 164, *165*
Sole, 146

Solid bridging, 147, 148
Solid-core doors, 87
Southern Building Code Congress
    International, 133
Specifications, 1
  air conditioning, 165
  plumbing, 158, *160*
Split system air conditioning, 164,
  *166*
Squares, 20, *21*
Stairs, 65
Stakes, 145
Standard brick, 151
Stone, cut, *85*
Structural details, 119–120, *121*
Stucco, symbol for, *85*
Supply air ductwork, 162, *162*
Supply air registers, 162, *163*
Supply water piping, 155–156
Survey plat, 133–134, *135*
Symbols, 45–46, *46, 47, 48, 49,* 85,
  *85*
  architectural, *192–193, 194*
  electrical, *195–196*
  on floor plans, 63
  HVAC, *199*
  pipe fitting, *198*
  plot plan, 136, *137, 193*
  plumbing, *197*

**T**

Tape measure, 27–28, *29,* 116, *116*
Terra cotta, symbol for, *85*
Tile
  dimensioning, 30
  symbol for, *193*
Title blocks, 4, *4,* 89
Tongue-and-groove planks, 102
Township, 133–134, *135*
Trade information, 145–166
Transverse sections, 99, 100, *101*
Trapezium, *21,* 22
Trapezoid, *21,* 22
Tread, stair, 66
Triangles, 7–8, *7,* 20, *21*
  acute, 20, *21*
  equilateral, 20, *21*
  isosceles, 20, *21*
  obtuse, 20, *21*
  right, 20, *21*
Trim
  exterior, 149
  interior, 149
Trim details, 120, 122, *123*
T-squares, 7, *7*
Two-story house, floor plans of,
    66–69, *69*
Typical detail, 99–100

**U**

Unit controls, and forced warm air heating, 162
Utility easement, 136

**V**

Vent piping, 158

**W**

Walers, 146
Walks, 147
Wall detail, 99–100, *100*
Wall elevations, interior, 117, *119*
Walls, 147–149, *149*
  exterior, 28–29, *29*
  masonry, 150, *150*
Warm air heating. *See* Forced warm air heating
Waste piping, 156–157, *158*
Water, potable, 155–156

Wayne Residence, 69–72, 89–90, 105–106, 122–124, 136–136
basement plan, 70
basement wall detail, 124
bathrooms, 70–71, 72
bedrooms, 72
breakfast area, 72
ceiling detail, 122
deck handrail detail, 122–123
dining room, 72
East Elevation, 90
entry, 71
family room, 70
fireplace detail, 123
foundation plan, 70
foundation wall, 70
garage, 72
hallway, 71–72
kitchen, 72
laundry, 72
living room, 72
longitudinal section, 106
North Elevation, 90
panned ceiling detail, 122
plot plan, 136

porch, 72
slab, 70
South Elevation, 89–90
stairway, 70
transverse section, 106
walkout detail, 124
West Elevation, 90
Welded wire fabric, *202*
Windows, 86, 87, *88*
  awning, 120, *123*
  details, 120, *123*
  dimensioning, 30, *30*
  double-hung, 120, *123*
  elevation view of, *48, 118*
  on floor plans, 66
  pictorial view of, *118*
  sectional detail view of, *118*
  symbols for, *194*
Window schedule, *87*, 90
Wood panel, symbol for, *85*
Wood siding, symbol for, *85*
Working drawings, 1–4
  defined, 1
Wythe, 105